MUSCLE AND BLOOD

MUSCLE
AND
BLOOD

RACHEL SCOTT

E. P. Dutton & Co., Inc. | New York | 1974

Published simultaneously in Canada by
Clarke, Irwin & Company Limited, Toronto and Vancouver
ISBN: 0-525-16150-3

Library of Congress Cataloging in Publication Data
Scott, Rachel, 1947-
    Muscle and blood.
    1. Industrial hygiene—United States.   I. Title.
HD7654.S35   1974   614.8'52   74-9862

To Vincent Kelly

CONTENTS

Go reckon our dead by the forges red
And the factories where we spin.
If blood be the price of your cursed wealth
Good God! We have paid it in.

—Anonymous (1908)

ACKNOWLEDGMENTS

Many, many people aided in the work that went into this book—
hundreds of workers, as well as union officials, reporters, health
professionals, government officials, and corporate officials. I can't
begin to name them all, but must thank Tony Mazzocchi and Steve
Wodka of the OCAW, George Perkel of the TWUA, Dick Sirginson, Dr. Thomas Mancuso, and Dr. Michael Utidjian for their
continuing assistance.

I want to thank Jim Boyd and the Fund for Investigative
Journalism, who provided money for researching chapters on the
Sunshine mine disaster and the Industrial Health Foundation.
Thanks to Marjorie Fine and the Abelard Foundation, who pro-
vided funds to the Center for Community Change, who aided my
work. Thanks to the Ford Foundation for financing a good por-
tion of the research and writing of this book, and especially to
Basil Whiting, who recognized the importance of the subject.

Thanks also to Tom Bethell, who provided me with a nearly
free office when I needed it, as well as an education in the politics
of coal-mine health and safety; to Pat Furgurson, who loaned me

his Blue Ridge mountain retreat, and to the MacDowell Colony, where I spent six weeks while at work on the manuscript.

Thanks to Mike Curtis at *The Atlantic Monthly,* who provided encouragement as well as valuable editorial advice during work on the Sunshine chapter which first appeared in that magazine.

I am grateful to David Obst who in September 1970 first introduced me to Tom Congdon, then an editor at Harper & Row, now editor-in-chief of adult trade books at E. P. Dutton.

Much credit belongs to Tom Congdon, my editor and friend, who believed in this book from the day it was proposed, and to the ideals it stands for, and who provided his support and enthusiasm as well as his many excellent editorial talents.

Thinking back on the years that have gone into this project, I realize how many roommates, friends, relatives, have wished me well and aided me as they were able, in what is, by its nature, a lonely task. Those people know who they are, and I thank them.

Rachel Scott
Brownsville, Maryland
February 19, 1974

MUSCLE AND BLOOD

MUSCLE AND BLOOD

Some people say a man is made out of mud.
A poor man's made out of muscle and blood.
Sixteen Tons

In 1906, in his classic book, *The Jungle,* Upton Sinclair wrote of the horrors of work in a Chicago slaughterhouse. "There were the men in the pickle-rooms, for instance . . . scarce a one of these that had not some spot of horror on his person. Let a man so much as scrape his finger pushing a truck in the pickle-rooms, and like as not he would have a sore that would put him out of the world; all the joints in his fingers might be eaten by the acid, one by one. Of the butchers and floormen, the beef-boners and trimmers, and all those who used knives, you could scarcely find a person who had the use of his thumb; time and time again the base of it had been slashed, till it was a mere lump of flesh against which the man pressed the knife to hold it . . . There were men who worked in the cooking-rooms, in the midst of steam and sickening odors, by artificial light, in these rooms the germs of tuberculosis might live for two years, but the supply is renewed every hour. There were the beef-luggers, who carried two-hundred-pound quarters into the refrigerator-cars; this was a fearful kind of work, that began at four o'clock in the morning, and that wore out the most powerful men in a few years. There were those who worked in the chilling-rooms, and whose special disease was rheumatism; the time-limit that a man could work in the chilling-rooms was said to be five years . . . There were those who made the tins for the canned-meat, and their hands, too, were a maze of cuts, and any cut might cause blood-poisoning; some worked at the stamping-machines, and it was very seldom that one could work long at these at the pace that was set, and not give out and forget himself, and have a part of his hand chopped off . . . Worst of any, however, were the fertilizer-men, and those who served in the rendering rooms. Those could not be shown to the visitor— for the odor of the fertilizer-men would scare any ordinary visitor at a hundred yards, and as for the other men, who worked in tank-rooms full of steam, and in which there were open vats upon the level of the floor, their peculiar trouble was that they fell into the vats; and when they were fished out, there was never enough of them left to be worth exhibiting."

Sinclair wrote later that in writing *The Jungle,* he was not thinking of the issue that made the book famous, the scandal of impure food. "I wished to frighten a country by a picture of what its industrial masters were doing to their victims," he said, "and entirely by chance I stumbled on another discovery—what they were doing to the meat supply of the civilized world . . . I failed in my original purpose."

The country, to this day, remains unfrightened. Most Americans believe—as they have been told—that the sweatshop factories with their unhealthy and dangerous conditions disappeared along with the sixty-hour week and child labor. On the contrary, present conditions in many thousands of American working places rival those Sinclair described. As a nation we have worshipped progress and profits, made gods of science and industry, blindly ignoring the evidence around us that they were destroying us— and not just our environment, our resources, but our people. Workers die daily in explosions and fires, are mangled by machinery, deafened by industrial clangor, and driven to the breaking point by harassment and the command to work at a dangerous pace. Hundreds of thousands of men and women—human beings with families and hopes and bodies as sensitive to pain as any person's—are poisoned at work by fumes and solvents and suffocated by lung-filling dusts. Yet, ignorant as primitive tribesmen of the human results of a burgeoning technology, most of them die quietly, their families accepting deceptive diagnoses such as heart disease, cancer, emphysema.

This book is about the continuing carnage which can be found hidden just beyond the most modern factory façade of shining steel and brick. To the age-old problems—the dirt, disease, and injuries of Upton Sinclair's day—even more serious perils have been added. Since World War II countless new chemicals have been developed, and employers have rushed to put the new chemicals to use, with little regard for their toxic properties. The effect has been human devastation. The 1972 "President's Report on Occupational Safety and Health" made a terrifying and probably conservative estimate of the toll: "There may be as many as 100,000 deaths per year from occupationally caused diseases," said the report, and "at least 390,000 new cases of disabling occupational disease each year."

In working-class families, these statistics would not seem incredible; everyone who labors knows someone who has been sickened or mutilated on the job. In their neighborhoods, the disabled may actually be seen dragging along the street, blighted, often young, workers whose livelihood is lost. But the great number of comfortable Americans, many of whom are now several generations beyond blue collar, have turned their backs on all that pain and death. Working in air-conditioned offices and living in stratified suburbs, they are oblivious to the plight of those they left behind. They can elude the appalling connection between their own prosperity and the human sacrifice upon which it is based. And so working people are left to the benevolence of owners and managers.

"We've tried over the last twenty years to talk safety to employers and get them to be good," Ray McClure, an industrial hygienist for the federal government, told me, "but as long as it costs money, they're not going to do it. It takes money for maintenance, it takes money for safety and health. I don't think it's fair to have a slogan like 'Safety pays.' It doesn't. In most plants it just doesn't. It doesn't turn a profit."

It became clear to me during three years of visiting America's working places that neither the government nor the unions nor the conscience of employers was protecting working men and women from the cruelest and most cynical use. Shortly after I began my research, a federal law was passed which for the first time was to regulate health and safety conditions throughout industry. The law promised much, but the hopes of working people, and the fears of employers, were not to be realized. The law would have little effect. As the following pages will show, some men are still permitted to increase their power and wealth by destroying the lives of their fellow human beings.

BERYLLIUM

"They Always Said It Wouldn't Hurt Us"

Hazleton is a former coal town in the heart of the eastern Pennsylvania anthracite region. The town—"Pennsylvania's highest city"—sprawls across the gently sloping Spring Mountain, in the Appalachians. As I drove down Church Street, past the Hazleton Public Library, the shopping center, and the new motels, heading north toward Buck Mountain and beyond to the Conyngham Valley and Sugarloaf Mountain, I thought how pleasant it once must have been in Hazleton—before strip miners changed the forested ridges and valleys to deep barren gullies and black mounds of debris—coal, rock, slate, and shale; what the miners call "gob piles." In this desolate wasteland nothing grows, nothing lives. Here where all seasons are colorless, hunters often find dead rabbits by the creek bed, poisoned by the black acid waters.

"As it came from the hand of God," wrote H. C. Bradsby in

his 1893 *History of Luzerne County,* "it was lovely to look upon. The keen-eyed pioneer beheld it, said it was good and here he would stick down his Jacob's staff and dwell forever. He heeded but little the obstructions that confronted him on every hand. The heavy forests that perpetually shaded the ground; the fierce and hungry wild beasts in the constant search of living prey; the gliding serpents spotted with deadly beauty, the countless birds of song and plumage and game; the fish disporting themselves and shining in the mountain brooks and in the beautiful blue river."

So it must have been in the early 1830s when a hunter digging for a groundhog found coal on Spring Mountain. Coal mining became a small industry in the area and remained small until several decades later when Tench Coxe, a wealthy Pennsylvanian, heard of the discovery of coal at Mauch Creek near Hazleton and bought eighty thousand acres, land that held some of the richest deposits of hard coal in the country. Coxe Brothers and Company started mining in 1865, wrote Bradsby, and in time became "one of the largest coal producers of any private house in the world." By 1890, Coxe Brothers was producing one and a half million tons of coal a year.

Tench Coxe was no ordinary coal baron. He was the great-grandson of Daniel Coxe, physician to King Charles II of England and later, in colonial America, proprietor of large tracts of land given him by the king in West Jersey, Carolana (sic), New York, and Pennsylvania. Tench Coxe's grandfather, also named Tench Coxe, was a wealthy manufacturer who served as Assistant Secretary of the Treasury under Alexander Hamilton, and, according to Bradsby, held the distinction of owning the first copy of Adam Smith's *Wealth of Nations* ever brought to this country.

Bradsby described Coxe as a man who would "spare no pains in improving the condition of the workingman in his employ" and as the "unconquerable champion of the rights of all." That was fortunate for his workers, for from Bradsby's description of a strike at the Coxe mine in Drifton, near Hazleton, it is hard to imagine what conditions must have been for men laboring under less gracious management. During the 1870s, Bradsby traveled incognito to the Drifton mines to see what the laborers had to say about Coxe. He talked to one worker, "a recently crippled laborer, who was just able to be out and was carrying a badly injured arm in a sling, who told of a strike of a few years ago; said that the miners at Drifton were ordered out and had to obey. They had an interview with Mr. Coxe and he frankly told them what would be the outcome; that they could not drive him; that he could afford to stop all work at Drifton far better than they could afford to be idle; that in the end they would have to go to work at probably less wages; that he could live if his property at Drifton was all at the bottom of a Noah's flood, etc.

"The men mostly knew that all he told them was the truth," said Bradsby, "but they had to obey orders, and after six months of idleness and all its consequent suffering were glad to resume work at less wages."

The strike was no doubt the bitter "long strike" of the winter of 1874–75 which idled mines throughout the hard-coal fields and ended in defeat for the miners and their union—then the Workingmen's Benevolent Association. The miner's lot in Luzerne County was as dangerous as it was difficult. Bradsby wrote of an 1869 disaster in the Avondale mine near Wilkes-Barre in which one hundred and ten miners were killed. Fire started in a furnace at the bottom of the mine shaft and flashed upward, destroying the only means of escape from the mine. The trapped miners died of suffocation from the smoke and fumes before rescuers could reach them.

When the mine could finally be entered, three days after the fire, said another historian, "Fathers and sons were found locked in each other's arms. Some of the dead were kneeling in the attitude of prayer; some lay on the ground with their faces down-

ward; some appeared to have fallen while walking. All knew that
the insidious influence of the surcharged atmosphere would soon
cause death."

The disaster focused the attention of the nation on the miners'
working conditions. Miners throughout the six-county hard-coal
region demonstrated spontaneously in Scranton, Hazleton, Mauch
Chunk, Bloomsburg, Sunbury, and Pottsville. A coroner's jury
concluded that the men might have lived if there had been a second
outlet to the mine. The Pennsylvania legislature, which in 1869
had rejected legislation for a comprehensive mine safety act, in
1870 after the Avondale disaster unanimously approved a similar
bill—the first for the state and the first mine safety act in the
country.

The mine disasters continued. In 1871 twenty workers died in
a fire at the West Pittston mine and seventeen died in an explosion
at the Eagle Shaft colliery in the same town. In 1885 forty-six
miners were killed in three separate disasters in Luzerne County
and in 1891 nine miners died in the Spring Mountain Colliery at
Jeanesville, just south of Hazleton. "So frequent are fatalities reported," wrote Bradsby, "that until one reflects how many people
are delving in the mines he is apt to conclude that life here is
precarious."

From the time of the Avondale explosion in 1869 (when the
Interior Department's Bureau of Mines first started keeping count)
through World War II, seven hundred and fifty men died in Luzerne County mine disasters; altogether, about fifteen thousand
men have died in mine accidents in the six-county Pennsylvania
anthracite region. The mine fatalities were reduced, finally, but
not by new compassionate mine owners. The mining itself was
giving out. By the end of war most of the coal mines near
Hazleton were depleted. In 1955 Hurricane Diana struck, flooding the few remaining underground mines. Most were never reopened.

After the underground coal was gone, the descendants of Tench

Coxe leased surface rights on the land to strip miners. Surface coal is usually less desirable than underground coal because of its high sulfur content, but it is cheaper to mine. The strippers ripped away the surface land with giant earth-moving machines and scooped out the coal underneath until that too was gone, and all that remained were wasted land and wasted men: many thousands of miners who after years in the dusty underground shafts were disabled by the hard-coal dust disease anthracosilicosis, once called "miner's asthma" and now nicknamed "black lung." So many of Hazleton's men were crippled by the disease that it became a leading cause of death in the town, trailing behind only heart disease and cancer.

With a dying coal industry, Hazleton suffered high unemployment. In 1956, seven thousand were unemployed in a town of thirty-three thousand—a rate of twenty-one percent for the area. Worried Hazleton businessmen organized "Can Do," an industrial development d.ive, raised three quarters of a million dollars from local businesses and banks and began construction of a 750-acre industrial park, with free land and ready-built industrial shells, roads, and sewer and water systems.

The first major industry to come to Hazleton was the Beryllium Corporation.* The corporation's main product, beryllium, was considered an exciting new metal in the 1950s. Almost all of it was sold to the Atomic Energy Commission, and so Hazleton boosters could claim that the town had made the jump from the coal age to the atomic age.

Beryllium does not occur in its element form in nature and was not discovered until the end of the eighteenth century. It is found naturally in many minerals, but principally as a component of beryl, a hexagonal glassy-looking crystal, the purest forms of which are the emerald and the aquamarine. In the Middle Ages, beryl was

---

* The company has changed its name to Kawecki Berylco Industries. But the workers use the old name, and that is the name used in this chapter.

thought to have mystical power. It could be used, according to Beryllium Corporation literature, "to foretell the future and review the past, to detect thieves and forewarn of death, and was supposed to have power over evil spirits that could be made to do the wearer's will by suitable incantations. Beryl rendered the owner cheerful, preserved and heightened conjugal love, and cured diseases of the throat and jaws and disorders of the mind."

When scientists were finally able to extract beryllium from the beryl ore in the early 1900s they discovered that the metal had properties almost as remarkable as those ascribed to beryl. Because of its low atomic weight, it was found to be lighter than aluminum yet stronger than steel, and more heat resistant. But its use in industry was limited by its cost. The ore must be "hand cobbed"—the crystals sorted manually from the rock in which they are found. The known deposits of the ore are rare and generally distant—in Argentina, Brazil, South Africa, India; only a few deposits have been found in the United States. Extraction of the metal proved costly too—the ore yields only eighty pounds of beryllium for every ton of ore. Thus this expensive metal was not used significantly in the United States until the 1930s. It was then that the lamp industry discovered that beryllium oxide had the curious property of fluorescing when exposed to ultraviolet light, and manufacturers began using it as a coating inside the glass tubes of fluorescent and neon lights.

In the late forties and early fifties, the Atomic Energy Commission foresaw increasing needs for beryllium, which looked promising as a "moderator" to speed the release of atomic energy in reactors. Water could also be used as a moderator (and is much cheaper) but beryllium saves space, so it was valuable in weapons (it was used in the Hiroshima bomb) and some thought it might make possible a small reactor for atomic-powered jets. The aircraft were proposed as part of a Polaris nuclear defense system—atomic-powered aircraft which, unlike conventional planes, wouldn't need to refuel but would be in the air continuously, able to retaliate with nuclear missiles in the event that enemy missiles destroyed U.S. ground defense systems.

To insure that the rare and costly metal would be available, the AEC awarded long-term contracts to the two beryllium-producing companies, the Beryllium Corporation and the Brush Beryllium Company, to build refineries which could produce high-quality beryllium. Rather than expanding its plant in Reading, Pennsylvania, the Beryllium Corporation decided to build a new factory.

"Beryllium was looking for a new site, and they had looked all over the eastern seaboard," said Dr. Edgar Dessen, president of Hazleton's booster organization, Can Do. "We had this worthless old railroad roundhouse, built during World War I. It was my job to sell them a bill of goods about coming to Hazleton." Dessen did his job; the company was delighted with the location.

The Beryllium refinery is at the end of a narrow blacktop road which winds through two miles of stripped mountainside east of town to a point beside Hazle Creek in the valley north of Spring Mountain.

Once a roundhouse, built in 1918 for repairing and servicing locomotives, the refinery is a huge half-circle, three stories high, dominated by a towering smokestack.

The Hazleton *Standard-Speaker* reported in 1957 that "Hazleton's cavernous strippings . . . were welcome sights to Beryllium Corporation officials as they decided to locate their new nuclear plant here. The unsightly strippings—ideal for waste disposal—coupled with a large male labor pool, isolation and ample utilities —were the deciding factors as Hazleton suddenly was whisked into the atomic age." A company official told the *Standard-Speaker* that Beryllium was particularly delighted that "on the site was ample sandy soil for waste disposal and a convenient polluted stream." Can Do gave the Beryllium Corporation seventy-three acres of land in addition to the roundhouse. Beryllium bought another two thousand acres from the Tench Coxe estate, the Hazle

Brook Coal Company, and the East Sugarloaf Coal Company.

Apparently few people wondered why the company was so particular about its location. If they did, the *Standard-Speaker* made no note of it. Hazleton men were glad to get the high-paying jobs. "Beryllium was the place to get into," one Hazleton worker told me. "The hourly rate wasn't exceptional at the time, but they were working seven days a week. Overtime—you could work twenty-four hours a day if you wanted. After forty hours, everything was premium time. If a guy wanted to put in the hours he could really make a buck."

As it happened, the strange new metal was far more dangerous to the workers than the coal dust that gave their fathers miner's asthma. Indeed, by that time, beryllium dust and fumes were considered to be a thousand times more toxic than coal dust. Exposure to beryllium dust, as to coal dust, can disable the lungs, causing coughing, chest pain, and shortness of breath. But beryllium disease is also systemic, affecting the heart, the liver, and other organs.

The hazards of beryllium were only too well known at the company's headquarters plant in Reading, just forty miles from Hazleton. A study of the plant by state health officials revealed forty-eight cases of beryllium disease by 1959. Eighteen of the victims were beryllium refinery workers, but nineteen were people who never entered the plant—mostly women whose main exposure to the dust was merely when they washed their husbands' work clothes. Twenty-one others were people who simply lived in Reading. Of these, according to a 1970 report by the Public Health Service, were "several individuals who lived several miles away from the plant and had tended graves across the street from the beryllium refinery." By 1963, seventy-five cases had been documented and at least thirty-one people had died.

These deaths might have been prevented had scientists and government officials in this country heeded European reports of the disease written some thirty years earlier. The toxicity of beryllium was recognized as early as 1933 in Germany and 1936 in Russia. By 1942, German scientists had reported that three workers had died from exposure to beryllium compounds.

"Why these reports were ignored in the United States is uncertain," said Dr. Harriet Hardy, who has since studied beryllium disease extensively, "but heavy responsibility rests with Public Health Service Bulletin 181, *The Toxicology of Beryllium,* published in 1943. Mention is made of some of the European reports. Then, after a description of a limited number of small-animal studies, the first line of the summary reads, 'The foregoing investigation indicates that beryllium is of itself not toxic.' "

Because of the official nature of the report, it was particularly valuable to the industry, and they seized upon it. The following year, three scientists, citing the PHS report, concluded that the epidemic of illness which was even at that time hitting workers at the Reading plant could not be due to beryllium. They blamed high fluoride concentrations for the disease (fluorides are released during the refining of beryllium) and they recommended that fluoride levels be reduced.

The first cases of beryllium poisoning in the United States appeared in the 1930s primarily among women working in the fluorescent lamp industry in Salem, Massachusetts. At first physicians thought the workers had tuberculosis. Dr. Hardy and Dr. Lloyd Tepper later wrote that when tuberculosis had been ruled out, doctors thought the patients had sarcoidosis, "although it was considered odd that all fourteen patients whose cases had been reported by 1944 had worked in the same industry."

Sarcoidosis (which is also called Boeck's sarcoid) is practically indistinguishable from beryllium disease. The main difference, according to Hardy and Tepper, is that people occasionally recover from sarcoidosis, a relatively benign disease, while chronic beryllium disease is incurable and has "not uncommonly been associated with a relentless down-hill course." The mortality rate for beryllium disease is thirty percent.

Scientists do not completely understand how beryllium attacks the body, but they do know that the most serious damage is caused by changes in the lung tissue, changes that impede the lungs' ability to pass oxygen to the blood. As the lung tissue grows useless, the right side of the heart pumps harder and harder to circulate more blood to the lungs. The resulting strain on the heart can cause "cor pulmonale," failure of the right side of the heart, the leading cause of death to victims of beryllium disease. Beryllium is apparently also carried by the bloodstream to vital body organs, such as the liver, heart, spleen, and muscle tissue, where it sometimes causes damage. It also interferes with the activity of certain enzyme systems.

The Salem epidemic was soon linked to the illness of other workers who handled beryllium. One doctor noticed the similarity of "Salem-sarcoid," as the Massachusetts cases were called, to the illness of six foundry workers who had worked with beryllium copper alloys.

Dr. Harriet Hardy began investigating the illness of Salem lamp workers in 1945 and has been studying beryllium disease ever since. Now in her sixties and semi-retired, she is the foremost authority on the disease and one of the most highly regarded scientists in the field of occupational medicine. She is a short, unpretentious woman, equally candid in her remarks whether speaking privately or delivering a speech. (A colleague says Dr. Hardy's grandfather once advised her to "say what you think and then you won't have to remember what you said.") A perfectionist in her own work, she will not tolerate mistakes from others. She mistrusts union leaders and reporters, to whom she is often inaccessible, and has even less regard for industrialists, with whom she is usually at odds. "Anybody in the beryllium industry," she once said, "is a rascal until proved otherwise."

In 1951 she established the Beryllium Case Registry at Massachusetts General Hospital, which by 1972 contained more than eight hundred carefully documented case histories of beryllium disease. But the eight hundred were merely the proven cases. She believed there were, conservatively, about twenty-five hundred cases, three times as many as had been reported.

During the forties, Dr. Hardy and other scientists studying beryllium disease met fierce resistance from the lamp industry. "Trade secrecy," she said, "was the reason given for refusing state authorities permission to study beryllium-using operations . . . With rare exception, industry and insurance companies withhold data on occupational disease . . . Figures on populations at risk are either inaccurate or not to be had."

Scientific knowledge was confused by some—in Dr. Hardy's polite term—"less well-meaning people" who continued to maintain that beryllium was harmless. *Product Engineering,* for example, in 1949 published an editorial entitled "Beryllium's Toxicity Is Largely Myth." (Dr. Hardy later remarked that the article was worked up by "one of the Pennsylvania alloy companies writing to explain why people like Harriet Hardy ought to be taken out to sea and drowned—and I have the lungs and livers from some of the men in that shop!") Another article of similar philosophy, "Taking the Brrr Out of Beryllium," appeared in *Industrial Medicine and Surgery* the same year.

The cases, however, Dr. Hardy said, kept "tumbling in." In 1947 the Saranac Symposium in New York established that "most if not all United States beryllium industries were suffering worker illness." In 1948 a committee appointed by the Atomic Energy Commission and headed by Dr. Hardy recommended that safe levels of beryllium dust and fumes inside factories be considered to be two micrograms (or two millionths of a gram) per cubic meter of air. Thus beryllium was established to be far more dangerous than such toxic metals as lead, mercury, arsenic, and cadmium —the standards for these heavy metals allow as much as 100 to 500 micrograms of metal per cubic meter of air.

In 1949 the new standards for beryllium were adopted by the AEC. The lamp industry, threatened by public panic about the hazards of beryllium in the fluorescent tubes and about growing numbers of workers suffering from beryllium disease, and facing the possibility of stiff new regulations, voluntarily decided to stop using the metal.

Cases of the disease continued to appear during the next decade, caused by exposures before 1949—symptoms may not appear for ten years after the last exposure to beryllium dusts and in a few cases have not appeared until twenty-five years later. In 1959 Hardy and Tepper were able to report that the two-microgram standard seemed to be safe for long-term exposure. "There have been no reports of disease at this level of exposure," they wrote, "during the ten years it has been used." At last the disease seemed to be under control.

During the sixties, the AEC's enthusiasm for beryllium waned— the Defense Department scrapped plans for nuclear-powered aircraft in 1961—but the amazing metal won new markets in aerospace. Boeing selected it, because it has high resistance to strain, for guidance system parts in the Boeing 747 jumbo jet. Because it is light, it was used in Lockheed's giant C-5A jet transport to reduce weight in the wheel brakes. Apollo space craft were built with beryllium structural parts and beryllium housings were developed for equipment used on the moon by Apollo astronauts.

Other industrial uses were developed. Beryllium copper alloys, much stronger and harder than copper and brass alloy, replaced these traditional materials in contacts and connectors in IBM computers and made possible the tiny microswitches which Honeywell manufactured by the billions for use in aircraft, vending machines, and machine tool control. The alloy made possible a smaller and more reliable temperature control mechanism for household appliances. "Any respectable refrigerator will probably have a beryllium copper control," said James Butler, special assistant to the president of Beryllium. "The same is true of electric ranges." Beryllium copper molds became popular in the plastics industry for producing intricate detail in plastic dolls, toys, and simulated wood TV cabinets and other furniture. By 1969 beryllium refiners were supplying beryllium metals, alloys, and ceramics to at least eight thousand industrial customers.

With the widespread use of beryllium in industry came new epidemics of beryllium disease. In December 1969, Dr. Hardy wrote to Dr. Marcus Key, then chief of the Bureau of Occupational Medicine at the Department of Health, Education and Welfare, that she "had hoped . . . to give up the missionary activities associated with my knowledge of beryllium poisoning and its control." However, she said, new cases continued to come in. "This week" she wrote, "has been especially full of reports from doctors and members of industry of cases of illness associated with various operations using beryllium copper alloys . . . This letter is stimulated by a call from a local medical examiner yesterday, reporting an autopsy he had just done on a relatively young man who clearly died of beryllium poisoning associated with investment casting of a beryllium copper alloy. This just should not happen. An additional episode which falls in the same category is a small 'epidemic' in a relatively small machine shop where four of sixteen machinists are said to have beryllium poisoning."

The Registry listed thirteen cases of the disease in 1970 and twelve cases in 1971. It is likely that the reported cases for these years represent only a fraction of the total number of cases. The unsuspecting physician can easily misdiagnose the disease—and often has, according to a 1969 mortality study by Dr. Thomas Mancuso and Dr. Anas Amin El-Attar. Mancuso and El-Attar checked death certificates of forty-three patients who had medically confirmed cases of beryllium disease. In only eleven cases did the physician note the presence of the disease. In thirty-two cases—seventy-four percent—the disease was not detected by the examining physician, and death was recorded as due to other causes—usually some form of heart disease.

Curiously, most of the cases in the late sixties and early seventies seemed to be coming from the secondary manufacturers, rather than the beryllium refineries. By 1970, Dr. Hardy had received reports for the Registry of only three cases of the disease from the Kawecki Berylco plant in Hazleton.

When the Beryllium Corporation built its Hazleton plant, town leaders knew there were certain risks, according to Dr. Edgar Dessen, a local radiologist. Dessen was Chamber of Commerce president at the time, and he had been a member of the company's board of directors since Beryllium came to Hazleton. "But," Dessen said, "the plant was, one—in an isolated area, and, two—built with all the engineering knowledge available. We were aware there was some danger, but I don't think nearly the danger that all our coal miners have been through with uncontrolled dust."

Soon after it began operations, the Hazleton plant failed to meet the AEC health and safety guidelines and was shut down. Dr. Donald M. Ross, chief of health and safety for the AEC, said the Beryllium Corporation was at that time "a Neanderthal company, which refused to admit for years and years, although its own doctor was discovering cases, that they had beryllium disease. And they had help in the state director of occupational safety at the time, who said when asked that he was not appointed to drive industry out of Pennsylvania."

The company installed new equipment and reopened several months later. After 1963, when the AEC terminated its contract with the company, and AEC standards no longer applied, said Dr. Ross, plant conditions apparently deteriorated. Ross accompanied an inspector from the Labor Department's bureau of labor standards on a plant inspection in October 1970. The inspector, Ray McClure, who was senior industrial hygienist for the bureau, cited several serious violations, mainly to do with poor housekeeping: "Pipes and overhead surfaces were dirty; material removed from dust collectors on the roof was not properly disposed of." He ordered the company to eliminate the open dump and to decontaminate waste going to the dump. "They were throwing old duct work and stuff out there that had a lot of beryllium dust in it," McClure told me. He also ordered the company to eliminate hand screening of beryllium, a procedure which exposed workers to hazardous amounts of pure beryllium dust. McClure took no air samples for beryllium, however, because, he explained, "Sometimes, frankly, they make everything seem all right. They knew we were coming. We can't get in there without notifying them ahead of time because of these classified contracts." (Several months earlier a Labor Department inspector had been denied entrance to the plant on the grounds that he didn't have "Q" clearance for AEC projects.)

Dr. Ross agreed with McClure's report by and large. "You know, the longer I stay in this business," he told me, "the more cynical I get. The feeling was, it seemed to me, that being accountable to somebody [as the plant had been to the AEC prior to 1963] had a kind of salutary effect. Their procedures, policies were not all that bad. It's just that they weren't following them. You know, unless there is somebody on your back—if you need to save some money, the place you can save it without too many people knowing about it is in health and safety."

And until about 1970, not many people seemed to know—or care. As conditions at Beryllium continued to deteriorate, workers became alarmed. Many of the workers found they were constantly coughing and short of breath. Others had been told their lung X-rays showed spots—spots which they knew could be a first sign of disease. The company actually admitted to three cases of beryllium disease. Two more workers filed for workmen's compensation, claiming beryllium disease, and local doctors reported still more cases.

The case that worried workers the most was probably that of Clarence Culp. Culp had symptoms that suggested beryllium disease—shortness of breath, cough, and extreme weight loss—though beryllium disease was never proved. One fellow worker said that before Culp died he was so thin "you could have touched your thumb and index finger together around his ankle."

Robert Ferdinand, another Beryllium employee, remembered Culp vividly. "It was pitiful," he said, "watching that guy come to work and standing there like a sick sparrow. Honest to God, it was pitiful how skinny he was—he'd come in so skinny with this big coat on and stand by the furnace just trying to keep warm. And we told him, 'Don't worry, we'll do the work.' I could see myself standing there the same way and I thought, 'No, that can't happen.' "

Ferdinand has beryllium disease now, and he thinks of Culp's death and is afraid. "When it gets dark," he said, "that's when I dread. I get so much pain—all over—my legs, my arms, my back, my neck, my head. It's hard to describe. Jesus, they're gnawing, grabbing. They just pull like I'm being pulled apart. Then a lot of times I get blurred vision and I can't see. I get dizzy a lot. And when I get these coughing spells, I spit up blood. And I get chills and sweats. I sweat something terrible sometimes. Water just pours off of me. And then I get the chills."

The first time I called on the Ferdinands, Robert Ferdinand was out. Kathleen Ferdinand was in the kitchen frying potato pancakes for lunch. She is a naturally attractive woman and neither uses nor needs makeup. She wore the pink housecoat that she wears most of the time, except on the rare occasions when she leaves the house.

"We got married the year I got out of high school," she said. "Robbie was out of the service, tending bar for his father. I worked in the kitchen and he took care of the bar for two years, because his father was sick and couldn't do it, and no one else would. Then he went with the boilermakers. He traveled all over, different states, to put big boilers up. He started to work at Beryllium in October 1957, only a few months after it was started. He was one of the first to be hired. And my first girl, Gail, was born in '55 and my second, Robert, in 1959. Then Jackie, and then I had my two girls.

"Everybody we knew that worked at Beryllium, they all expressed the same feeling of being tired, you know, itching. Right from the beginning they don't feel good. I don't know. My husband should be coming back soon, I hope; he can tell you better than I. But the people we knew, they'd talk about the place, how dirty it was.

"But we didn't realize then that it was toxic. I never *worried* about it. But if I had known that it was toxic, believe me, he would never have worked there. And his eyes always bothered him. Always bothered him. Redness around the eyes." She laughed. "He says, still today, when the wind is blowing from Beryllium, that his eyes—he really swears it—that it's from beryllium."

We could hear someone at the front door. *"Daddy's home!"* cried Jackie, Ferdinand's ten-year-old son, running into the kitchen and out again. Several minutes passed before Ferdinand appeared and sat down at the table. He leaned over, supporting himself with his forearms on the table, breathing in deep gasps, sucking the air through his mouth.

"Hi, how did it go?" his wife asked.

"Okay, I'm tired," he said, ending the statement with an apologetic laugh. "I'll get my wind."

His wife looked at him. "He walks up the stairs and he gets . . ." She shook her head; her voice was tight. "I . . . I was saying . . . last night, he woke me up when"—Ferdinand was coughing now, a dry rapid cough—"when the pain pills wore off . . . he gets a lot of pain . . . he was moaning and all, until they took effect again and he fell asleep. I couldn't go to sleep anymore. I was through."

Ferdinand was still coughing. She reached for his arm and squeezed it, then withdrew her hand self-consciously. "I don't know," she said. "If . . . if we could, I'd like to try another climate, or something. I'll go to Australia if we have to."

"We're going to the Fiji Islands," he said gently, smiling at her. "I'm going to find a nice island someplace where nobody's ever been. A store to buy food in. Her and the kids 'n somebody to talk to, I guess. That's all, I guess." He sat gazing at her. A large handsome man with soft dark hair and a serious face except for his long, almost comical nose, he looked younger than his forty-three years.

"We try to keep the kids' spirits up," said Mrs. Ferdinand, "because they get downhearted, you know, from being in so much. The only thing that bothers them is, their daddy's sick, and they want him to get better. And they can't understand why . . . why we have to do without so much just because he got sick. Why is it that because he got sick we can't afford to do anything or buy things. They do a pretty good job though. They're good kids."

"Yeah," agreed Ferdinand, "we got good kids."

"Jackie came home with a valentine on Valentine's Day, and on top he had, 'Daddy, I hope you get better.' " She laughed nervously.

Ferdinand said, "It could have been anything. But—'I hope you get better . . .' "

"Yeah," she said. "I hope you get better. On a valentine!"

"I hope you're saving it. Are you?"

She nodded.

"Good."

Neither spoke for a moment. We could hear the television in the next room—the rattling, tuneless music of children's cartoons.

Ferdinand had been "healthy as a trout," as his doctor put it, when he started working for the Beryllium Corporation. After a few weeks in the plant, however, he began to cough. That first winter he fell and hurt his back.

"The snow was pretty deep," said Ferdinand. "I had to go from the plant where I worked over to the warehouse to get some parts. When I come out of the door I slid on the steps. The steps weren't shoveled off—there was all snow—and I didn't know anything until I hit the bottom. There was eight or ten steps I fell down."

Ferdinand went to the plant nurse the next day. "She said, 'You didn't fall far enough to hurt yourself,' she said. 'Don't worry about it.' " But his back got "worse and worse" and finally he went to his family doctor. Doctors told him he had a herniated disc, which normally would call for an operation, but that was considered too risky with Ferdinand's cough, which by then was persistent. He lay in bed for six months before he was able to return to work.

Ferdinand was a conscientious worker at Beryllium, and he assumed that the company, in turn, was protecting him from hazardous exposure to beryllium. "They always said it wouldn't hurt us. They say your tolerable level of exposure to an eight-hour period is, what—two micrograms? That was for an eight-hour period. I worked eighteen, twenty—I don't know how many hours overtime. Not only me, everybody did. I think I worked as much overtime as I did straight time."

Sometime in the mid-sixties—Ferdinand doesn't remember precisely when—the company medical director, Dr. Edward Henson, called him in. "He said I had spots on the lungs on my X-rays. Said, 'Don't worry about them, you probably had bronchitis one time or another.' " Ferdinand believed him. "Like I am, I trust everybody. I didn't let it bother me."

A few years later, in 1968, Ferdinand was working on a blocked acid line when sulfuric acid overflowed from a tank, filling the area with poisonous fumes. "I was outside at the time. I tried to run in and see if I could do anything. My respirator was upstairs where I was working. I went up the steps and I couldn't see too good. So I went back and got fresh air out the door and I went back in to see if I could help put water in the pit to wash it down. I already saw a guy there. He had a respirator on and he was flushing the area down. I went outside, the fumes just followed me right outside and I couldn't get away. I couldn't get any fresh air at all. I had to run all the way to the end of the building to get some fresh air. After that I got nauseous and sick to the stomach.

"After that, every day kept getting worse and worse. One night, two weeks later—I was working on the roof that night—I was so sick I was throwing up. Coughing so hard I couldn't keep anything down. So I went in and said to the foreman, 'I'm too sick to work. Give me a pass. I've got to go home.' That was April 19, 1968. That was the last time I worked."

The Beryllium Corporation agreed to pay compensation—sixty dollars a week for the family of seven, the rate set by state law (Ferdinand's average weekly wage had been two hundred dollars) for "irritative tracheo-bronchitis." The company checks often were several weeks late. "My mother would give us food," said Mrs. Ferdinand. "We didn't have anything to eat half the time."

For the five weeks before Christmas that year, no checks had arrived. The Ferdinands called and pleaded with company officials, but the money did not come. On Christmas Eve, Ferdinand called the plant personnel manager, Carmen Fornataro.

"I was down in the basement, putting clothes in the washer and I heard this loud voice," said Mrs. Ferdinand. "I couldn't imagine what happened—he was just sitting there quiet all evening—I didn't know he was going to do it. He picked up the receiver and called Fornataro's home. What did you say to him?" she asked her husband. " 'I hope you're having a nice Christmas'?"

"Yeah," said Ferdinand. "I didn't have any money to buy my kids anything and I was just down in the dumps. I tried to call Mr. Velten [the plant manager] and I couldn't get any answer. I would have called the president of the company, but he wouldn't give me his number—Carmen wouldn't give me his number."

The check arrived the day after Christmas.

That January, Beryllium sent Ferdinand to Dr. H. S. Van Ordstrand at the Cleveland Clinic in Cleveland for tests. Dr. Van Ordstrand, like Dr. Hardy, has studied beryllium disease since the 1940s. The difference, I was told by James Butler, an official of Beryllium, is that "people like Van Ordstrand have one view of this thing and Dr. Hardy has another. Dr. Hardy looks at this as a dose-response disease. [In other words, that the illness is related to the degree of exposure to the dust.] Now the Van Ordstrand approach is that beryllium is a disease of individual susceptibility. We're inclined to go along with the latter theory."

According to the "individual susceptibility" theory, a worker could be exposed to large amounts of beryllium dust and still not get sick, unless he was "allergic" to it. Dr. Hardy told me the idea of individual susceptibility "is a lot of damn foolishness. The Registry data tells us that as controls began to be used, the numbers of cases fell off dramatically. And we know—or are satisfied to say—that there is *no such thing* as a beryllium poisoning in which there is anything magic about its occurrence. No sensitivity, nothing curious. If the amount is held at the safe level, nobody is going to get sick."

Van Ordstrand reported to the company medical director, Dr. Henson, that Ferdinand's physical examination was "essentially negative." A pulmonary function test, he said, was "nondiagnostic and compatible with his being a bit overweight"—225 pounds at a height of five feet, eleven and a half inches. Beryllium patch tests were "negative except for a very slight reaction to a one percent solution of beryllium fluoride. I would interpret this as either his own sensitivity to the fluoride rather than the beryllium part of the solution or of it confirming his history of previous beryllium fluoride dermatitis.

"Our examination here on Mr. Ferdinand," the doctor concluded, "showed no evidence for any chronic berylliosis. [Berylliosis is another name for beryllium disease.] I am really unable to explain his respiratory symptoms except for the possibility of their being in the nature of an asthmatic bronchitis at times . . . I would think that if they were in the nature of an occupational allergy, that he should have become free of them when off work."

Dr. Henson called Ferdinand to his office to tell him the results. "He told me I could go back to work," Ferdinand said. "Maybe most of the guys might have gone back to work—would have took their word for it, but when you *know* damn well something happens and you're *sure* of it, you have to stick with it." Ferdinand went to his own doctor, who arranged for a lung biopsy the same week.

Beryllium discontinued compensation payments, and the Ferdinand family went on welfare. It wasn't until four months later that Ferdinand received the biopsy report confirming what the Ferdinand's feared—that he did indeed have beryllium disease.

Dr. Herman Feissner has been the Ferdinands' family doctor since Robert Ferdinand was a boy. An old-style country doctor, he has lived in the Hazleton area all his life and has been practicing medicine since 1935. During the day he is full-time physician for the White Haven School and Hospital for the mentally retarded. Nights and weekends he devotes to his private practice in Freeland. He has patients with black lung—"hundreds of them"—and now Ferdinand and possibly one or two others with beryllium disease.

He described Ferdinand, before his illness, as "a very stable, responsible individual. A well-rounded average American man. The entire family was always an industrious, hard-working family. I don't know any that laughed at the law," he said. "I think they took the law seriously and lived that way. His father was a coal miner and later a tavern owner. His mother was a typical coal-cracker housewife—very rarely got out of the apron. Some of the family were caretakers for the Coxes. As the late Mr. Daniel Coxe remarked to me, 'Salt of the earth.' "

Dr. Feissner is an independent man, not afraid of controversy, yet he hesitates to talk about himself, and he related the story of his first encounter with Beryllium with reluctance. When the plant first opened, he said, the company invited the Hazleton branch of the Luzerne County Medical Society to tour the plant. "The company provided transportation for us. The bus took us to the plant and then—I thought—it was to take us back to Hazleton. But after the tour I was informed they would continue to the country club for dinner." Dr. Feissner and another physician got off the bus. "We requested to leave because we didn't want to be good-will ambassadors for a company we didn't know anything about." And the more he knows, he said, the less good-will he feels.

When Ferdinand left work in 1968, Feissner was convinced that Ferdinand's symptoms indicated beryllium disease. The symptoms were classic. Loss of appetite, dry cough, irritability, chest pain, shortness of breath, weight loss, and exaggerated reflexes. He said the biopsy findings showed 138 milligrams of beryllium per gram of lung tissue and a granulomatosis change in the lung tissue which is characteristic of the disease. The lung tissue feels grainy, as if it is full of little seeds.

"The mere presence of beryllium in a tissue is no indication of a disease," said Dr. Feissner, "but when you have a granulomatosis change—well, it's pretty obvious that if there's a thief hiding behind the sofa you're not going to blame the disarray of the room on the good housekeeper. Same thing here."

Dr. Victor Greco, the chest surgeon who performed Ferdinand's lung biopsy, agreed with Feissner. "You can find beryllium levels in a miner's lungs that are increased without having berylliosis. But if all these other things fit, the company can always say they have something else, but when you have the X ray, the history of exposure, the lung biopsy, the proof now is overwhelming. Who are you kidding?"

Yet company officials were not overwhelmed and they did not agree with Ferdinand's doctors. Ferdinand retained a lawyer, Bartel E. Ecker, and filed for Social Security and state workmen's compensation benefits. The Social Security hearing was held as scheduled, in August 1969, but four months passed without a decision. In December Mrs. Ferdinand wrote her Congressman, Representative Daniel Flood, to find out what was causing the delay. The answer shocked the Ferdinands. "The record is being held open pending receipt of additional evidence from Mr. Ferdinand's attorney," Charles M. Erisman, acting director of the Social Security Administration Bureau of Hearings and Appeals, wrote in a letter which Congressman Flood forwarded to the Ferdinands.

"The hearing examiner office has contacted the attorney several times, requesting the additional evidence. When this material is received, the hearing examiner will complete his evaluation of the entire record and issue a decision."

Whatever Ecker's reasons were for delaying the decision, he didn't bother to mention them to the Ferdinands. "Every time I called to ask how it was going," said Mrs. Ferdinand, "he just said, 'It looks good.' " When the case was finally decided in February 1970, Ecker's share of the disability benefits—set by law at twenty-five percent of the amount accrued from time of disability to first payment—was $941, about $300 more than he would have received if the award had been made a month after the hearing date.

Even before the Ferdinands had discovered the cause for the delay they had decided to find another lawyer. Ferdinand was especially upset that Ecker had told him, "What do you care what they say you're disabled from, as long as you're disabled?" From December through March 1970, they were unable to find a lawyer willing to take their case. One lawyer promised to talk to a friend who was a member of the local bar association. The lawyer later called back Ferdinand to say he had been advised to tell Ferdinand to "remain with Mr. Ecker"—as it happened, a law partner of Ecker's was president of the local bar—and for that advice he charged the Ferdinands fifty dollars. Another lawyer Mrs. Ferdinand called asked her to hold the line for a moment and never came back. "I held the phone for half an hour before I hung up," she wrote in a letter to Congressman Flood. The couple talked to the attorney a second time in his office after the Social Security claim had been settled. The attorney "gave us a few minutes of his time, wrote down some of the information, and said he would get in touch with us," said Mrs. Ferdinand. "We never heard from him again. At one point my husband just sat and cried. I didn't know what to do. I finally wrote to Mr. Ecker and asked him if he would continue with our cases . . . He waited almost two weeks to reply. He refused."

Eventually, through assistance of the local labor council, the Ferdinands found another attorney, but they were disappointed with him too after he pressed the family to accept an "amicable" settlement of forty-five dollars a week for three hundred and sixty weeks.

The partial settlement was proposed, I was later told by Oscar Brown, attorney for Beryllium and a member of the board of directors, "because the doctors say he wasn't permanently disabled. Plus the fact that he could take a job and that would supplement his earnings." Which would have been fine, I replied, if Ferdinand had been able to work. "Yes, we thought so," said Brown, missing the sarcasm, "and his lawyer thought so."

The Social Security Administration didn't think so; it was already paying him $418 a month for total disability. Ferdinand didn't think so either. He flatly rejected the proposal. In the fall of 1971, Ferdinand was still waiting to go to hearing on his compensation case. By this time, through the assistance of the Oil, Chemical and Atomic Workers, which had recently replaced District 50 of the United Mine Workers as the workers' union at Beryllium, he found another lawyer. And the union sent him to see Dr. Hardy in Boston.

"She didn't hem-haw around," said Ferdinand later. "Either you have it or you don't." He had it. "I was doubtful before. Till I saw her I was always hoping when I went to see her she'd say I didn't have it. But she said I did, so I just have to live with it." After three years of medical controversy, the diagnosis was a relief.

The ordeal affected the Ferdinands' political views profoundly. "I used to feel bad when I saw hippies," said Ferdinand. "That they were causing trouble they shouldn't. But now I think they have a reason to do a lot of things that they do. I can't see burning and looting and killing people. But they have a protest. They *notice* something wrong with our laws."

His wife nodded her agreement. "They notice something, yeah."

"It has to show up," he said. "And you find out in cases like this it shows up. There's something wrong. You have to go by laws, you just can't run around wild and free. But jeez, if your laws aren't run the right way, there's something wrong that's got to be straightened out somehow. I always had a respect for law—doctors, lawyers, priests. Not today. I don't know. Is everyone paid off? Is it just my imagination? Or what is it? I don't know."

The experience had taught the Ferdinands enough that Mrs. Ferdinand was able to write a sophisticated letter to her state senator:

Dear Sir:

I am writing to you in the hopes that it may serve as a particular in the eventual changing of some of our laws, for it seems everything works to the advantage of the employer.

Workmen's compensation, I am sure, has succeeded in plunging countless hard-working men and women into instant poverty when they became disabled, though it has probably proved to be a bonanza for the nation's profit-dripping insurance companies. The workmen's compensation process, with its mechanisms for stalling, must be partly to blame for the fact that occupational injuries are steadily on the rise. Coupled with the fact that when a worker is covered under workmen's compensation laws, [he loses his] right to sue for negligence. Thus the companies are free to allow hazardous conditions to exist without fear of any big court awards. I was very surprised to learn that when we finally do get compensation, all over 80% of his average monthly wage (with Social Security and compensation combined) *is taken back off him.* At a time in his life when he needs it the most. We had recently gotten a raise in Social Security. This came at a time when my husband was taking medication that required a new prescription every third day, at $10 a prescription. He also began taking oxygen at that time. Because of the raise, the DPA [state department of public welfare] could not help us with either. So you see the laws are quick to see that you don't get enough, even though you become sick because of a company's total disregard for health and safety, they are not answerable, and suffer no inconvenience . . .

I understand there is a new health and safety act that went into effect April 1971. The first meaningful job health legislation in the history of our nation. I commend those who helped pass it.

I cannot begin to describe the tragedy for the people (like my husband) for which this program arrived *too late.* No one has figured out what it really means to the man who becomes sick (through no fault of his own) in terms of what happens to his family, its finances, its hopes and plans, and its long-term security. My husband loses, I lose, and most of all my children, the next generation, loses . . .

It seems to me if labor and industry would put life and health before profit, it would be the solid foundation needed for both to prosper. So that others will not be assured a slow progressive industrial disease to be the wages of his work.

I'll close hoping something will be done to lift the financial burden for those who are ill . . .

Sincerely yours,
Mrs. Robert Ferdinand

Since Ferdinand's illness first raised the specter of a new epidemic of beryllium disease, more cases have been discovered. Ferdinand's cousin, John Curry, twenty-three, after only two months in the plant, developed the same symptoms Ferdinand had. Curry's doctors, Dr. Peter Saras and Dr. Greco, diagnosed beryllium disease. Curry, however, was not totally disabled, and thus, under Pennsylvania occupational disease laws, he was not eligible for workmen's compensation benefits. Yet with a record of lung disease, he would have trouble finding another job. Companies are reluctant to hire workers with health problems that could later cost them a compensation claim.

Dr. Greco told me about another patient of his who developed pulmonary disease after working at the beryllium plant. "Nobody would hire him with this history, and Beryllium wouldn't hire him back either. They sent him to Cleveland Clinic and they said all

he had was bronchial asthma. I tried to get them to either say he
had berylliosis and give him compensation, or hire him back. Finally when their hand was forced," he said, "they hired him back."

In 1971 the Pennsylvania Occupational Health Department
made a medical survey of the plant. Of a total of 336 workers who
were questioned, thirty-six had a history of "chemical pneumonitis." Nobody knows whether those cases were due to exposure to
fluorides which are given off in the refining process or whether
these were actually cases of acute beryllium disease, caused by
high exposure to beryllium fumes. "It's perfectly true," said Dr.
Hardy, "that acid salts—fluorides, nitrates, and so forth—are
capable of producing an irritation of the bronchial tubes which has
nothing to do with beryllium." She said sometimes a worker will
experience an early period of lung trouble which clears up and
then later he will develop beryllium disease and disability. "What
isn't clear," she said, "is whether the original episode is caused by
a gradual accumulation of small amounts of beryllium over a
period of years."

More alarming than the pneumonitis cases, however, were the
X-ray findings. Of the 336 workers surveyed, fifty-six, or sixteen
percent, had abnormal lung X-ray patterns. The number could
have been considerably higher if Ferdinand, Curry, and others no
longer in the plant were included.

The Thomas Hornack family, for example, did not appear in the
statistics, although it has been hard hit by the disease. Hornack,
the son of a coal miner—his father died of miner's asthma—was a
maintenance mechanic at Beryllium. His father-in-law, James
Tallerico, and a cousin, Peter Greco, once workers at the plant,
were among the three cases of the disease that the company agreed
to compensate for total disability. And Hornack's wife, Angeline
Hornack, may be the first "community case" of the disease in
Hazleton. Doctors were not sure whether she had beryllium disease or the illness that it so closely mimics, sarcoidosis.

In past days at other refineries, wives of workers were exposed to beryllium dust when they washed their husbands' contaminated work clothes. At Beryllium work clothes could not leave the plant. Workers had to shower before they picked up their own clothes to wear home. But workers thought they might still carry home dust.

"They have fall-out at the plant," said Thomas Hornack. "This stuff would fall over the cars in the parking lot and the parking lot is not paved. They paved all around the plant but where the worker parks his car they haven't done anything. They claim it's too expensive. In the summertime, it's hot, you leave your car windows down and dust gets in. And when you leave, everyone wants to get out in a hurry. They raise all kinds of dust and that gets in the car too." Hornack has had to replace his car windshield twice because of pitting from the chemical fall-out. "There must be three hundred fifty men working down there," he said, "and I'd say almost every one has had a new windshield put in."

Hornack's health may be in question too. For six months in 1961, as a victim of "chemical pneumonitis," he was restricted from certain areas of the plant. "I almost passed out," he said. "I could hardly breathe at all." Eventually the restriction was lifted, but ten years later, when I interviewed him, he was still suffering from chest pains and shortness of breath.

By 1969, Beryllium workers had become concerned enough about deteriorating plant conditions to take action. That was when they voted to replace their union, District 50 of the United Mine Workers—which one worker said "never stood behind our safety and health"—and voted in the Oil, Chemical and Atomic Workers, a union with a reputation of active concern for working conditions. The OCAW, alarmed by workers' tales of conditions at the plant, began to investigate.

The union assigned the task of looking into the complaints to Steve Wodka, a young legislative assistant in the OCAW's legislative office in Washington, D.C. Wodka visited the plant and met with Beryllium management. The initial encounter, Wodka told me, "was not the kind of confrontation you would expect. First they told a whole string of jokes. They ended up with a marijuana joke—something to take me off balance. I always find that disgusting, to have to laugh along. At some point it's going to get deadly serious."

Wodka found the plant conditions to be as bad as workers had complained. During the tour, he said, he was walking through the furnace area "when all of a sudden this guy runs out. He takes me up this tower. There at the second or third flight was this beryllium dust. You could see it filtering down through the stairs. Guys were climbing up and down it all day long. The decontamination crew is supposed to come along and clean it up, but it had been there for a week! We took Velky up there and asked him how can this be tolerated?" (Leonard Velky was the company industrial hygienist.)

Wodka found problems throughout the plant—poor ventilation and visible beryllium dust in some areas, fumes from toxic chemicals and acids in other areas. "We asked the company if they would give us the results of the monitoring. This would give us an indication of what the beryllium levels have been running inside the plant. Velky wouldn't release it to us. He said it was a policy of the company from the president on down never to release this kind of information."

Beryllium also refused to show Wodka results of a state survey of plant conditions. Nor could he get the report from the state. Edward Baier, director of the Pennsylvania division of occupational health, said the state's procedure was to release information only to the company. He claimed that the survey might jeopardize company trade secrets, a rationale that not even Beryllium had thought to suggest. Beryllium's reason for keeping the information secret, explained C. Dale Magnuson, industrial relations manager for Beryllium, was that "we don't feel the union is knowledgeable enough about the subject." Beryllium at first also rejected a union proposal that Dr. Hardy and her associates at MIT conduct engineering and medical surveys at the plant and, based on their findings, make recommendations for improving safeguards.

"We feel we're doing the job that is required," said Magnuson. "We don't see why it would help to have them come in and re-invent the wheel, so to speak."

Meanwhile, Wodka continued to pressure state officials to take action. Inspectors returned to the plant to do more air monitoring from March to May of 1971. The results, which this time were released to both the company and the union, indicated beryllium levels as high as 5,000 micrograms. Clearly the workers were ex-posed to an unhealthy, possibly deadly, atmosphere.

Beryllium officials scoffed. The test results could not be ac-curate, they said. Yes they could, the state maintained. "We've been inspecting the beryllium plant at Hazleton from the first day they began to remodel the roundhouse," said Joseph Memolo, regional industrial hygienist for the state. "The plant has had high results from time to time, and we have made recommendations as to what should be done to control the exposure to beryllium. I imagine the plant has done their best, as far as I can see, to comply with our recommendations, but the more we sampled, it seems, the higher results we got until just last May—March through May—when we did our final sampling. Our results were extremely high. It appears to us there may be a condition there that warrants severe improvement."

Suddenly Beryllium agreed to the union's proposed survey, pos-sibly with hopes that the results would be less dramatic—after all, where could the dust count go but down? But when the scientists from MIT inspected the plant in October, their results simply con-firmed the state survey. The company took another blow from federal officials who made a third survey, cited the company for twenty-six safety violations and five "serious violations" for "ex-cessive beryllium concentrations in work place areas" and fined it a less-than-staggering $928. Listed among the top thousand corporations in the country, with net sales in 1970 of $61.4 million and profits of $1.5 million, Beryllium could pay the fine out of petty cash.

Still, after a year of constant pressure from workers and their international union, after at least four critical surveys and much unfavorable publicity, conditions remained unchanged. Throughout the haggling and the debate, the refinery's three hundred and fifty employees, afraid for their lives but held by their need to support their families, continued to work in a poisonous environment. The likely result—new epidemics of beryllium disease—may not appear for another ten years. By then it will be too late and the deaths will be brushed off as the unfortunate product of past mistakes.

METALS I

Making Steel and Killing Men

A sunny midsummer day. There was such a thing sometimes, even in Coketown.

Seen from a distance in such weather, Coketown lay
shrouded in a haze of its own, which appeared impervious
to the sun's rays. You only knew the town was there, because you knew there could have been no such sulky blotch
upon the prospect without a town.

. . . The wonder was, it was there at all. It had been
ruined so often, that it was amazing how it had borne so
many shocks. Surely there never was such fragile china-
ware as that of which the millers of Coketown were made.
Handle them ever so lightly, and they fell to pieces with
such ease that you might suspect them of having been
flawed before. They were ruined, when they were required

to send labouring children to school; they were ruined, when inspectors were appointed to look into their works; they were ruined, when such inspectors considered it doubtful whether they were quite justified in chopping people up with their machinery; they were utterly undone, when it was hinted that perhaps they :eed not always make quite so much smoke . . . Whenever a Coketowner felt he was ill-used—that is to say, whenever he was not left entirely alone, and it was proposed to hold him accountable for the consequences of any of his acts—he was sure to come out with the awful menace, that he would "sooner pitch his property into the Atlantic." This had terrified the Home Secretary within an inch of his life, on several occasions.

However, the Coketowners were so patriotic after all, that they never had pitched their property into the Atlantic yet, but, on the contrary, had been kind enough to take mighty good care of it. So there it was, in the haze yonder; and it increased and multiplied.

—from *Hard Times,*
Charles Dickens

The Syracuse Foundry in Syracuse, New York, was a small operation, employing about a hundred men. The foundry earned some brief notoriety in 1970 during Adam Walinsky's campaign for state attorney general. Walinsky held a press conference in the foundry parking lot to publicize poor working conditions there. During the press conference the president of the Syracuse Foundry gave emphasis to Walinsky's remarks by physically assaulting a reporter.

The president, Lionel Grossman, was no more interested in publicity in 1971 when I called to ask for a tour than he was in 1970. He refused even an interview. So I sneaked in. I was amazed at what I saw. The molten iron had only just been poured into molds on the dirt floor. With no ventilation, the smoke and fumes rose slowly from the floor, filling the building with a dense, sweet-smelling haze. Workers were visible only as shadows in the fog. With the smoke screen as protection against the president, I talked to workers for thirty minutes and left undetected.

I learned from the workers that the plant had recently been organized, although the company had yet to recognize the union. Several men, they said, had died from "tuberculosis" and others suffered from chronic headaches and breathlessness. Workers had complained to management about their chief problems—no ventilation, inadequate respirators, six showers for one hundred men. In the summer, there was extreme heat. In the winter it was cold in the plant, and the potbelly kerosene stoves scattered throughout the building, the only source of heat, were not vented to the outside; their fumes went directly into the air in the plant. But to all the complaints of the workers, management had turned a deaf ear.

A worker who was afraid to give his name told me why the men had to put up with the bad conditions. "Most of the guys is in debt," he said. "They owe a lot of money and they scared if they strike, they'll lose their homes and automobiles, and them that's renting, afraid they'll be throwed out or something of that sort, and that's why they keep going on. If they had their bills paid up and had enough money saved where they could live through a few weeks, I don't think there's a man in the shop wouldn't strike. It's a terrible place to work, and they don't do nothing about it. They laugh at everybody working here."

The foundry was no laughing matter. In the short time I was in the plant, the thick dust stained my shoes and permanently ringed my socks. My throat and nose were black with the dust and when I blew my nose the mucus was black too. The oily cloying foundry fumes saturated my skin, my hair, my clothing, and lingered for days, and nothing would remove them. When I stopped in a medical library in Syracuse after visiting the plant, people moved away from me to avoid the smell.

At the library I had hoped to find some explanation of hazards to foundry workers—mortality studies, perhaps, which would shed some light on whether foundry employees showed higher incidences of diseases commonly associated with dusts and fumes, such as heart disease, respiratory disease, or lung cancer. I found French studies, Italian studies, German studies, and a few British studies, but in the American literature, nothing. How little progress we have made since 1910 when Alice Hamilton, the first industrial physician in the United States, began her investigation of "the dangerous trades," as she called them. She wrote about her first experiences with industrial diseases in her autobiography, *Exploring the Dangerous Trades:*

> It was also my experience at Hull House that aroused my interest in industrial diseases. Living in a working-class quarter, coming in contact with laborers and their wives, I could not fail to hear talk of the dangers that working men face, of cases of carbon-monoxide gassing in the great steel mills, of painters disabled by lead palsy, of pneumonia and rheumatism among the men in the stockyards . . .
>
> There was a striking occurrence about this time in Chicago which brought vividly before me the unprotected, helpless state of working men who were held responsible for their own safety.
>
> A group of men were sent out in a tug to one of Chicago's pumping stations in Lake Michigan and left there while the tug returned to shore. A fire broke out on the tiny island and could not be controlled. The men had the choice between burning to death and drowning, and before rescue could arrive most of them were drowned. The contracting company, which employed them, generously paid the funeral expenses, and nobody expected them to do more. Widows and orphans must turn to the county agent or private charity—that was the accepted way, back in the dark ages of the early twentieth century. William Hard, then a young college graduate living at Northwestern Settlement, wrote of this incident with a fiery pen, contrasting the treatment of wives and children of these men whose death was caused by negligence with

the treatment they would have received in Germany. His article and a copy of Sir Thomas Oliver's *Dangerous Trades,* which came into my hands just then, sent me to the . . . library to read everything I could find on the dangers to industrial workers, and what could be done to protect them. But it was all German, or British, Austrian, Dutch, Swiss, even Italian or Spanish—everything but American. In those countries industrial medicine was a recognized branch of the medical sciences, in my own country it did not exist.

In spite of my failure at the library, I could not believe there were no studies of present-day American foundries. But calls to federal and state health officials confirmed that, indeed, no one knew how foundry workers may be reacting to their often hazardous environment.

Irving Davis, chief of the Michigan division of occupational health, told me in 1973 that foundry workers may be exposed not only to deadly silica dust but also noise, heat stress, carbon monoxide, various resins such as phenol formaldehyde, hydrocarbon, and coal tar pitch. He said he believed Michigan's foundries, which employed some forty thousand workers, had shown "gradual improvement" since 1965 when the state made its last comprehensive dust survey. Whether the improvement had been enough, clearly no one knew. "There's room for a lot of work here," Davis said.

In spite of a total absence of scientific studies of foundries, the silica hazard has generally been acknowledged to be a severe one. A. J. Kaimala, an industrial hygienist for the Detroit Bureau of Industrial Hygiene, put it bluntly. "Show me a person that worked in a plant with silica thirty years and I'll show you a person with silicosis."

Three foundries in Muskegon, Michigan, employing thirty-five hundred to four thousand men have for years produced about four hundred new cases of silicosis annually. Ten percent of the entire work force, in other words, was disabled every year. One large law firm represented most of the workers' compensation claims until 1972, when the firm was dissolved, and senior member Ben Marcus said the firm grossed one million dollars annually in lawyers' fees from foundry claims. His partner, Jerry McCroskey, told reporter Jon Lowell of *Newsweek* that the average Muskegon foundry worker is a sure bet to contract silicosis if he stays on the job long enough. "It takes twelve years in dirty jobs, twenty years in relatively clean jobs," he said.

The foundries may not have done much to protect their employees, but they have acted vigorously to protect themselves from liability for the human damage they cause. In 1966 the foundry industry lobbied successfully in Michigan to create a state silicosis fund. Under the fund, foundries were made liable for only the first $12,500 awarded each disabled worker. When a worker is awarded more, the balance comes from a special state fund financed by every employer in Michigan who pays a compensation premium. Cleaning up the foundries would involve better ventilation and a larger maintenance crew. That, of course, would cost money—more money than it costs the foundries to cripple their entire work force every ten years.

Muckrakers of the early 1900s described brutal conditions in the heavy industry of their day. In 1907, William Hard, the student referred to by Alice Hamilton, wrote in an article called "Making Steel and Killing Men" that during a single year—1906—at a United States Steel Corporation plant in Chicago, 46 workers were killed, 184 temporarily disabled, and 368 permanently disabled.

"The operating men who manage the Illinois Steel Company are human beings," he wrote. "They do not wish to commit either murder or suicide. But steel is War. It is also Dividends. All the operating men in South Chicago, from William A. Field down to the lowest 'Huniak' who now sculls the ladles that Mr. Field used to scull, are bound, hand and foot, by the desire to produce more steel this month than was ever before produced in South Chicago. The figures that indicate products and profits are the only figures handled and scrutinized by the members of the board of directors of the United States Steel Corporation. Steel is War."

Accident rates in steel mills and similar heavy industry have been reduced dramatically since 1906—due in part to reforming efforts of Hard, Upton Sinclair, Ida Tarbell, Dr. Alice Hamilton, and others of that day, and social legislation their reports inspired. Another period of social consciousness, the 1930s, and the rise of industrial unions, brought more reforms. After passage of the Walsh-Healey Public Contracts Act in 1936, which regulated work conditions in plants under federal contract, injury rates began to drop. They continued to fall for the next twenty years. Then in 1958, accident rates throughout industry began to climb again until by the end of the sixties they had increased twenty percent. A U.S. Labor Department report, dated 1966–67, said: "Substantial rises in injury frequency occurred in industries manufacturing primary metals, fabricated metals and machinery . . . Contributing to the rise in the frequency of injuries has been the growth in manufacturing employment particularly in durable goods industries, a high rate of new hires, and heavy overtime." Also contributing was the continuation of profit-hunger Hard described.

In the early spring of 1971, Bethlehem Steel's Lackawanna plant in Buffalo, New York, ranked as second largest of the corporation's steel mills and fourth largest in the United States. Bethlehem agreed, through spokesman Frank Banker, to give me a tour. "You'll see steel made and you'll see it made safely," he said.

I met Banker at the company's offices in downtown Buffalo. As he drove me to the plant, he told me that the mill was operating with a work force of fifteen thousand people and encompassed a broad array of operations in some three hundred buildings on eighteen hundred acres of land bordering Lake Erie. Banker said the plant produced 6.7 million tons of raw steel yearly, though it was scheduled for a cutback in production later that year when the aging open hearth furnaces would be permanently shut down.

"This plant is unusual in the Bethlehem system," he continued, "because it's a lake port." As we drove through the plant gate, the lake was hidden from view by immense heaps of limestone and beyond the limestone, hills of powdered soft coal, the essential ingredients, along with iron ore and scrap metal, for the making of steel.

"They say Lake Erie is dead," Banker said. He frowned and shook his head. "It isn't dead. Why, some of the best fishing on the lake is right here off the slag heap!"

The first stop beyond the limestone and coal were the coke ovens, a long row of connected narrow black steel structures, each oven about fifteen feet high, two or three feet wide, and forty feet deep, from front to back. In the first stage an oven is "charged"— a funneled vessel on top of the ovens pours the powdered soft coal into the empty oven, a dusty operation which sends a black smoke pouring out over the yards. The coal is cooked until it becomes coke, a hard dense fuel used in turn for charging the steel furnaces. Every few minutes an oven is ready either for charging or for the coke push. In the coke push, a manually operated machine on one side removes the oven door, then rams the orange-hot coke out the opposite side into an open car which moves on tracks beside the ovens. As the coke is pushed out, so are orange flames and a heavy black smoke. The rail car carries the hot coke into an enclosed quenching tower, and immediately a huge column of steam spews from the overhead stack.

The steam shadowed the sun as we watched. When I looked back down at my notes, the page was covered with black dust. We circled around the ovens to the downwind side and the air was thick with dust and fumes.

Banker pointed out that workers were wearing protective gear— face shields, asbestos coats, gloves. "We try to get them to wear respirators where they are needed, but they refuse to. This is where the human element comes in."

Later I talked to Mike Reilley, a Bethlehem worker who had spent some time at the coke ovens. "It's a very dirty, dusty, filthy operation," said Reilley. "They give you a respirator. It stops the bigger chunks in the air, but the smoke and gases go right through. You're exposed to a great deal of heat. You're working on top of furnaces with a lot of smoke in the air and a lot of times you can't see where you're going. When you come out of there you're dehydrated and really filthy. In the summer in many of these departments, heat exhaustion is a common thing.

"Five years ago," said Reilley, "a gang of about six guys were working along and pretty soon one guy keeled over, and the next thing they knew they all had to be carried out." Reilley worked only two days on top of the ovens before he couldn't stand it any longer. "They sent me to Dr. Mehnert," he said. Dr. Robert Mehnert was a company doctor. "He just told me we had guys who worked here sixty, seventy years and they worked happily. He told me there were no gases up there, that the conditions were safe and adequate. It wasn't that the coke ovens were bad, it was just that we were used to better conditions. No attempt to make a medical evaluation, just a heart-to-heart talk. 'You don't need to breathe, you don't need to see—you'll pick that up.' That's their attitude. The workers on the coke ovens are ninety percent black. So you have the poorest people in these jobs with the least education. Dr. Mehnert never took the attitude, 'Well, let's see if these conditions are bad and what we can do to change them.' It's very common for men to complain. He knew what the conditions were like by heart before I went in."

Coke oven workers risk excessive exposure to coal tars. Coal tars have been known to produce cancer since 1775, when a British scientist observed that London chimney sweeps suffered commonly from scrotal cancer. Coke oven workers, as it happens, also develop cancer of the scrotum—at a rate five times that of the general population. They also develop cancers of the lung, bladder, and kidneys at rates greater than the general population. The most recent data available on American coke oven workers is from a mortality study by C. K. Redmond and J. W. Lloyd. Comparing coke workers to other steelworkers in a Pennsylvania steel mill, they found that workers who had been employed five or more years in the coke ovens died of lung cancer at a rate three and a half times that for all steelworkers. The men who work on top of the ovens seem to experience the most hazardous exposure; their lung cancer mortality rate is ten times that for all steelworkers. Other studies of coke oven workers have suggested a higher incidence of cancers of the larynx, nasal sinuses, pancreas, stomach, and blood-forming organs (or leukemia), according to a review of the literature published by the National Institute for Occupational Safety and Health.

After watching the coke ovens, Banker and I visited the basic oxygen furnaces—these in a tremendous building, eight stories high, the furnace platform four stories above the ground floor. Scrap iron is moved to these furnaces first by electromagnets, which load it into "scrap muckers"—vessels on tracks. These scrap muckers haul it across the floor, to a giant crane, twenty feet wide, with hooks fifteen feet long. The crane lifts the mucker up to the furnace platform. Another machine holds the mucker until the furnace, like a giant cement mixer, is tilted to receive the iron. Then the mucker is tilted too and the scrap iron slides into the furnace with an ear-splitting clatter and rumble. White flames and sparks fly up and out in a huge, blinding fireworks display. The charging takes about five minutes. After the charging the "oxygenated blow" begins. This takes about twenty minutes. Banker said the furnace produces 285 tons of steel at one time, in less than an hour.

The newly formed steel is poured into ingots, and the ingots are carried by rail to the soaking pits, where they are reheated to 2,300 degrees for rolling. We watched the red-hot ingots carried along horizontal rollers to vertical rollers where they were squeezed through, returned, flipped, then squeezed again. Their shape slowly changed from a chunky block to a long thin slab. An operator controls the rolling from an enclosed booth above the rolling floor.

Then the slabs are rolled to a second process, called "scarfing," which blasts the steel surface clean, releasing a murky yellow gas. The scarfing operator is also in an enclosed cab. Scarfing is accompanied by a long, loud screech, and Banker and I were forced to cover our ears, the sound was so piercing. The heat was ferocious. I had a headache now and my breathing was affected by the fumes.

Plant tours are seldom terribly revealing. It's easy for management guides to steer even the most practiced observer away from hazardous areas, and I was certainly not a practiced observer of steel mills, though I did notice, because I was looking for them, "crippled" railroad cars. Bethlehem operates its own transport system of 166 miles of track at the plant, and the railroads are a whole business by themselves. In a strike dispute in 1969 the United Transportation Union, representing Bethlehem railroad workers, charged that workers were forced "to handle crippled cars, on which such things as grab handles and steps are deficient." A brakeman quoted by the Buffalo *Evening News* said, "At night, you'll grab in the dark and not see that there isn't a step there, and lose a leg or be dragged under the moving car." I could see for myself that the situation had not been remedied. I saw a sign scrawled in yellow chalk on the end of one car, STEP LOOSE, and a yellow arrow pointing down toward the offending step. Handrails were missing and bent on other cars.

Bethlehem literature boasts of the company's health and safety programs. "Every conceivable effort is made to insure the safety and well-being of employees at the Lackawanna Plant," said a brochure. "Great emphasis is placed on the prevention of accidents. Instructions in, and demonstrations of, safety measures and practices are given, and working conditions are constantly inspected to reduce possible fire or accident hazards. Atmospheric tests are made daily to detect contamination that might adversely affect the health of employees. A plant clinic is equipped to treat injuries and emergency illnesses, and to make thorough physical examinations." But workers and local newsmen told a somewhat different story.

David Fanning, a reporter for WBEN-TV in Buffalo, covered a series of accidents at Bethlehem in March 1971. Late one Friday night a scaffold collapsed near the blast furnace and five men were injured. The next day a crane fell in the structural steel shipping yard; no one was killed. (The overhead cranes in steel yards are enormous and powerful. Their bodies arch high in the air and move over the heads of workers and materials below. When one falls, most likely from inadequate maintenance, workers below are all but doomed.) A third accident, an explosion in which one worker was hospitalized with burns, occurred the following Tuesday. Fanning said Bethlehem denied his requests to film the plant. "The night we ran the story," he said, "a guy who said he is a safety engineer at Bethlehem called and said if he ever got another job he'd come in and tell us some stuff."

Mike Reilley, the Bethlehem worker mentioned above, worked in the bar mill. He complained of noise which "will deafen you permanently—I've temporarily lost about twenty percent of my hearing." He spoke of the cold in the winter: "From the waist up you'll be freezing, and you'll be boiling from your waist down from the ovens." And he told me of the smoke produced when steel bars hot from rolling are stacked between layers of wood. "All this produces is a temporary eye irritation, but it's just one of those little things they're always putting at you—to make life a little better," he said sarcastically. "Everybody feels they're beat down to the ground. They're crazy and they drink all the time. Ninety percent of them are alcoholics. It's a job that really drives people nuts.

"They paint these inane signs on the wall, like THE WORKER WHO THINKS IS A SAFE WORKER. Any worker must have seen hundreds of them. That's about the extent of their safety—putting these signs up on the wall. I mean, they really don't give a damn.

They'll do anything to avoid calling a time-lost accident a time-lost accident. About a year ago this one guy, he got his leg messed up. Nothing real serious—normally he'd be off, like, five weeks. Man, they had him right back in there."

Perhaps the best evidence that making steel at Bethlehem is a hazardous business comes from Dr. Joseph M. Dziob, director of Lackawanna's medical service. A story on the medical service in the Buffalo *Evening News* reported that the clinic, to which most of the Bethlehem injured are brought, recorded 58,956 emergency patient visits for treatment in 1969, including 37,642 new cases and 21,314 repeat visits. Ninety-seven patients were sent to hospitals for a total of 1,014 hospital days. "In comparison," said the story, "Buffalo's emergency hospital, probably one of the busiest emergency centers in western New York, treated 15,919 emergency-room patients in 1969, with 1,135 eventually admitted." Figures like these make it hard to accept the adage steel spokesmen are fond of repeating—that a worker is safer on the job than he is in his own home. The statistics to support that claim—provided by the National Safety Council—are based on lost-time accidents. Dr. Dziob admitted that Bethlehem held lost-time accidents to a minimum by the simple device of keeping injured men on the job, if at all possible.

"It would not be uncommon to me," he said, "to have a man come here with a broken leg which I would set under local anesthesia, put in a cast, and then maybe send him back to his job where he would be given selected work like sitting in a chair with his leg up. This helps keep down the lost-time hours."

Ken Bellet, a chief steward at the Bethlehem plant, told me of accidents fully as horrible as those of William Hard's day. "A couple of weeks ago I saw a man wrapped up in a steel bar," he said. "A cobble [an irregular slab of red-hot steel], when it flies off the rollers, loops in the air—it will wrap around anything. It wrapped right around the man. He was burned bad. I seen one accident where a piece of flying steel split a guy's head open and cauterized the wound in the same instant. A couple of weeks after I started working there a foreman was killed. A lift fell on his head and squashed him like a bug. One guy had a heart attack near the furnaces and died waiting for the car to take him to the dispensary. There was a man killed about a year ago—his sleeve caught in a round straightener and he got whipped to death. There's an Arab we call Snake. One day he had his arm on a pile of steel bars. The crane set a load down on his arm and his arm ended up like a waffle.

"Tony Balon got his fingernail ground off in the grinder. He went to the dispensary. They took bone clippers and clipped off his finger. They did get the doctor dismissed—and they gave Tony a hundred and fifty dollars for the finger that he lost.

"Another man was running an upset hammer—a big heavy hammer that runs on compressed air. A piece flew off and went right through his penis. He almost hemorrhaged to death waiting for the car to the dispensary—it took them over an hour to come. He's still loused up. He's working, but walking bowlegged over that thing. And he was always quite a ladies' man."

It is an unusual worker who doesn't have a half dozen stories to tell about his own experiences with injuries and difficult working conditions, and Bellet, who was thirty-five when I interviewed him in 1971, was no exception.

Bellet had been hired as a welder in 1967. "I worked eighteen months, put in a grievance one day, the next day they told me I was a laborer," he said. "They had grease barrels that weighed five hundred pounds. They'd have me roll them across the yard in the snow, by hand. Normally they were moved by truck. For almost a year I got almost every rotten job they had. I kept running to the steward to get them to end this and he'd do nothing. He was a drunkard and in a bad position. Finally I circulated a petition to get him fired as steward. Everyone except the chief steward signed it. The union refused to recognize it.

"The safety conditions they have there, mechanics carry a blue flag, a safety flag. They put it on the switch while they're working on a machine, warning them not to throw the switch. One time I was working on a straightener repairing a grease line. I had my flag on the main switch. I was reaching in between two rolls trying to get the grease line out. The foreman came over hollering to get that blue flag off the switch, it was holding up production. I told him if I didn't replace it the line roller would freeze and maybe a bar would fly off and hurt somebody. His comment to that was 'I don't give a good fuck,' and he pulled the flag off the switch and turned the machine up with my arm between the rolls.

"If a man gets hurt in the mills," said Bellet, "there's no first aid or emergency treatment on hand. The foreman has to call for a car to come get you. And it's three miles from the dispensary to the bar mills. The first week I worked there I smashed my fingers between two coupling irons—coupling irons weigh about a hundred and fifty pounds each. They swelled up like crazy. The skin was broken. I thought they were broke. I waited two hours for the car, and one and a half hours in the waiting room at the dispensary. I finally got to see a doctor. The doc told me to wash my hands in the sink. Then he X-rayed me. He told me the X-ray showed no broken bones, and gave me a slip to return to my regular work. He never bandaged it. The skin was broken. A week later I was back with an infected hand.

"The second time I was injured I was working near a welder and my eyes got flash burns. Again I had to wait hours for the car. I went to the dispensary, the doctor squirted some greasy stuff in my eyes and told me I should go home. I told the foreman I was going home. The foreman said, 'Why go home? Go down to the finishing end. Sit in the shanty and no one will know the difference. Why lose the day?'

"The next morning the foreman wrote me up for not doing anything all night long. Then they started putting me to work as an oiler. My hands broke out in a rash. They would itch like crazy. Then they would start to dry out, and when I would bend them they would crack and bleed. So I went to the dispensary. Dr. Mehnert looked at my hands. He asked me what was wrong. I said I got a rash on them and that I never got it before I was oiling. Mehnert said, 'You're a very nervous person. The perspiration glands on your hands causes them to break out. It has nothing to do with oiling.' So he gave me a slip to return to work.

"Mehnert had marked on the slip, 'non-occupational.' I said to him, 'Don't you usually test when a person has a rash to find out what the cause is?' He said, 'Don't tell me my business. I'm a doctor.' I told him that was debatable. I demanded that I be tested. So they sent me out to North and Franklin Street to have a Dr. Jordan look at me. Dr. Jordan looked at my hands, and he said, 'You're a very nervous person, you perspire a lot on your palms and you have a restriction on your perspiration glands that causes this. Of course we can't be sure without tests. But,' he said, 'in my opinion, that's what it is.' So he wrote me a letter to take back to Dr. Mehnert. It read, 'this illness is non-occupational.'

"I blew up. I said to Mehnert, 'The man told me he couldn't tell without tests. He didn't test me and the man concludes it's non-occupational. I demand to be tested.' Mehnert called me 'a fucking troublemaker.' Then the next day he sent me back to Dr. Jordan. He was visibly angry. He said to me, 'Who do you think you are?' He said, 'Myself and Dr. Mehnert have both confirmed this is non-occupational and you think you know more than the doctors.' I asked him was he going to test me, or was I sent to be lectured to. He said, 'I'll give you the test, but it's to no avail.'

"Meantime he had some guy from the safety department running around the bar mill taking little gobs of greases and caustic— he took some of that—it would eat paint off of steel. A couple of days later I was back at Jordan's. He was putting gobs of grease on my back. I seen the labels on them—I tried to tell him one of them was caustic. He said, 'We'll have to test you for everything you come in contact with.' Four days later he removed the tapes. The tape that held caustic left a burn mark on my back. One—

grease number 102—and another, a solvent, showed a reaction. I had a rash all over my back. The doctor said, 'Well, you did have a reaction!' He was really shaken over this. He wrote a letter, advised I never be given the position of oiler again and that I return to my regular work which was welding.

"I showed the letter to Mehnert. He took it, glanced at it, and handed it back to me. Said, 'Show it to your boss.' So I gave the letter to my boss. The boss gave me a pair of rubber gloves and told me to go back oiling. I still got the rash. The grease could get down into the rubber gloves. When you are greasing you get oil all over you, not just on your hands.

"A month ago there was a change in supervision. The new guy put me back on welding. I was welding on a mill stand, welding an oil dripper. The thing fell off and struck me on the knee. I told the boss I was going to see my doctor, that I was injured. He said he'd call for the car to take me to the dispensary. I told him I didn't choose to go to the dispensary. I chose to go to my doctor. He blew up and said, 'You always got to give people a hard time.' He made me wait in the office while he made five different calls to find out if I could do that. There's no posting anywhere in the dispensary telling the New York State compensation act—telling of the right to go to your own physician. I punched my card out. They docked me for the rest of the day. If I went to their doctor I wouldn't have been docked. The injury wasn't serious. I came in to work the next day. They told me they wouldn't allow me to work unless I'd go through the dispensary and get a complete physical. The Bethlehem compensation man just happened to be there and he grabbed me. He said, 'There's nothing wrong with the system. You're just causing a lot of confusion.' He asked me what I had against company doctors. I said there was a question either of competence or honesty. They always seemed to belittle an injury."

While we talked, Bellet gave me a tour of his neighborhood, the same one where he grew up, in his old wheezing, clanking car. His father was a garbage man, he said, and the family house was on an alley not far from Bresnahan's Hall. We drove by and the hall was still there but the house was gone—slum clearance. "It's a funny thing," he said. "When you live in a slum, you never think of it as a slum."

When he lost his first job as a meter reader, at the age of sixteen, for walking across a rich man's grass, his mother explained, "There's three classes of people—the upper, the middle, the lower. You're on the bottom and you ain't never going to get up."

Bellet could never accept his mother's words. When he was sixteen years old he read Jefferson's *Letters on Democracy*. "That book has caused me a whole lifetime of discontent," he said. "Jefferson said, 'It seems to me that the masses of men are born with saddles on their backs.' When I was a little kid going to grammar school, American history was a real thing to me, where you could hear the roar of cannons and the roll of drums. Here were people with a very radical idea. It took them a tremendous struggle and after all that struggle, to see it disintegrate to the mess there is here! Before I came to Bethlehem I had a job at Linde [a division of Union Carbide]. I was head welder there. I'd been in the hospital a few times for breathing problems. I was in for two solid months in 1965 with pneumonia. Then I got a job at Bethlehem. My wife begged me to keep my mouth shut about the breathing problems. We were starving. I was offered a foreman's job, but if you make a lot of money, what good is it if you have to sell out to do it? You buy a nice house and you can't hang a mirror on the wall. What the hell good is it? I got my high school equivalency and I've got thirty-five hours to go and then I'll have a degree in sociology. Then I'm going to leave those plants forever and become a social worker."

Bellet and his wife and five children lived in a second-story apartment crammed with dilapidated furniture. The neighborhood was forbidding. Though it was nighttime, Bellet sent his eight-year-old daughter down to the corner store to buy soft drinks for us. I would have been frightened to go myself, and I shuddered to think of the small girl skipping down the dark narrow steps of the tenement and out into the deserted streets. I hoped that Bellet would get his degree and get out.

Republic Steel's Buffalo mill, the smallest in the Republic system, was a dwarf compared to the Bethlehem plant. On a walking tour conducted regularly for the public, I saw most of the plant in two hours. Afterward I had pizza with two of the tour guides. They asked me what I thought of the plant.

"Some of those places we didn't go in looked pretty bad," I said. We had walked by several buildings which I could see at a glance were poorly ventilated and filled with fumes.

"That's why we don't go through there," said one of the guides.

The other added that the ventilation fan in the scarfing department had been broken for two months before it was repaired. "You probably didn't notice," he said, "but if you'd have looked to your left as we walked by the blooming and scarfing, you would have seen men scarfing by hand with acetylene torches." The fumes where they work are so thick, he said, that as you walk through you can see only a few feet ahead.

Men complain about the conditions, the first guide acknowledged, but most foremen, coming up from the ranks, figure that's as high as they are going to go. Afraid of losing their jobs, they ignore workers' complaints and actually laugh at the men.

After the tour we had seen a propaganda film produced by the American Iron and Steel Institute. As we watched scenes of men at work, a voice announced: "The work may look dangerous, but because of training and proper safeguards, this is one of the safest of all industries. As a matter of fact, steelworkers are safer in the job than at home. It takes rugged men to make steel, but today's steelworker must have education and knowledge, too. Ingots today, weighing fifteen to twenty tons, are easily shaped by men sitting in air-conditioned comfort."

My guides readily admitted that they didn't believe that film any more than I did. The first guide said he had told one of the blooming mill operators he ought to see the film—he'd enjoy hearing about the "air-conditioned cabs" he was supposed to be working in. They work in no such comfort. Earlier, during the tour, the operator had told me, "Whatever the temperature is outside, that's what it is in here." And with ingots 2,300 degrees Fahrenheit passing beneath his cab, that can be hot indeed.

An unusual operation at Republic, generally considered obsolete, was a manual bar mill operation. Milling is the final step in making steel. It is the process in which the red-hot steel is shaped to its final form, whether sheet, structural, pipe, or bar, by passing through a series of rollers. In one bar mill operation, a thin bar emerged from one series of rollers and a worker caught the red-hot steel with a long pair of tongs, whipped it around in a half circle, and threaded it between another, smaller set of rollers.

"Most people work down there four hours," a Republic worker told me, "then go off and don't show up for three or four days. You work twenty minutes on, thirty off, or thirty on, thirty off. Nobody can stand that heat for very long. Sometimes a piece is cold and won't bend. That's murder. It just keeps going straight and there's nothing you can do. But usually it's a glowing white turning orange when it gets to me." Fires are a constant danger. "Almost everyone has burns. It's the most common type of injury down there. There's fires almost all the time. My socks caught on fire twice. One guy, his entire suit caught on fire. He's still in the hospital. They don't think he's coming back."

Another hazard was the increasing use of unfamiliar chemicals, he said. "On some steel they put a chemical on it for galvanizing. The first time I smelled it, we were all over the rail puking our guts out. A lot of times the only thing I'm able to hold down is tea and a cookie."

Overhead cranes were a hazard at Republic too, the worker said. One fell because the operator overloaded it. "It fell, and me and this other guy almost said our last rites." Republic also

maintained its low lost-time accident rate by holding injured workers on the job, according to the worker. "You can't leave until
at least two hours before quitting time," he said, "unless you leave
in an ambulance or go to the morgue. They'll have you work with
broken bones. Just a few weeks ago they had one guy catching
[steel] with a broken hand."

Frank Palombaro, president of the Steelworkers Local 1743 at
Republic Steel, impressed me as a friendly, easygoing, and straightforward man. He expressed reservations about the federal Occupational Safety and Health Act, which he thought had a lot of
loopholes. The union didn't keep records on accidents, but Palombaro said the union's next contract "will have a big stress on
safety."

"Safety is an important factor, but there's an awful lot of people get hurt. I know my father worked in the steel plant forty
years as a crane operator. He worked over pickling vats. He contracted emphysema—whether that was work related, nobody
knows. But you just can't breathe all that dirt and fumes and
sulfur and—when you have leaded steel—lead without having
some effect.

"There's a lot of accidents that should be lost-time accidents,"
Palombaro said. "As long as a guy can walk in the plant, they'll
get him in there." Palombaro himself injured his arm, tearing the
ligaments in his left shoulder, several years ago while he was working as an electrician in the plant. "They tried to claim it was
arthritis, not an injury in the plant. I didn't lose any time, I just
did work where I didn't have to lift my arm. Now every time I
move it, it hurts." He also suffered deafness—from the noise, he
believed. "Some places it was constant noise, eight hours a day,
and when I went home I would still hear a ringing in my ears
almost until I went to bed. Now my wife sits across from me. She
says something, she says, 'Didn't you hear what I say?' "

The men themselves sometimes "forsake safety," he said, to
make production. "Most workers are on some kind of incentive

program. Every department has a different type of incentive. Foremen are also on incentive and top management have a bonus system. Under contract terms, a man can refuse to do the job because it's unsafe. The guy will holler but then go ahead and do it. Otherwise, he's going to lose money that day. It's going to affect his earnings. If it means shutting down a whole department, the other guys will shoot you." The result is that often rather than shutting machinery down for repairs, maintenance workers have to work while it is operating.

"The unions have strived to make things as easy as possible," he said. "It's a continuing battle for us. I don't think there are going to be any drastic changes. We're going to have to fight and fight and fight. It's a continuing battle."

METALS II

"Slow Death Is What You Call It"

". . . Now, you have heard a lot of talk about the work in our mills, no doubt. You have? Very good. I'll state the fact of it to you. It's the pleasantest work there is, and it's the lightest work there is, and it's the best-paid work there is. More than that, we couldn't improve the mills themselves, unless we laid down Turkey carpets on the floors. Which we're not a-going to do."

"Mr. Bounderby, perfectly right."

"Lastly," said Bounderby, "as to our Hands. There's not a Hand in this town, Sir, man, woman, or child, but has one ultimate object in life. That object is, to be fed on turtle soup and venison with a gold spoon. Now, they're not a-going—none of 'em—ever to be fed on turtle soup and venison with a gold spoon. And now you know the place."

—Mr. Bounderby, a wealthy manufacturer
of Coketown, from Dickens' *Hard Times*

I first became interested in the Montana metals refining industry in 1970, when I read an HEW summary report of pollution problems in East Helena caused by lead smelters operated by Anaconda and by American Smelting and Refining Company. Because of lead, arsenic, and cadmium pollutants from the two plants, ranchers have not been able to raise horses in the area since the early 1900s. Commercial rabbit growers frequently report stillborn litters. Henry Schroeder, M.D., a professor of physiology at Dartmouth Medical School's trace-element laboratory, wrote to a Montana scientist that he was "not at all surprised that . . . rabbits have dead litters. This, of course, is typical of lead poisoning. I would suspect that the East Helena mothers will also, in time."

I wondered then what effects the workers themselves must be experiencing since they are exposed to far more concentrated quantities of these pollutants. I called HEW's Bureau of Occupational Safety and Health, nicknamed "BOSH," to find out. I asked Pat Foley, the bureau's public relations officer—there was only one—for the reports of two studies mentioned in the HEW summary report. From the beginning she seemed doubtful that the reports were available, so I pressed for an explanation.

"Who made the study?"

"I think we—at their request—assisted the Montana State Health Department. I think there are certain confidentialities that they can insist that we keep—that we are not allowed to release."

"Do you know now if I can have that study?"

"No, this is what we're going to have to find out. And I hope it doesn't take the general counsel to decide."

"Who is the general counsel?"

"Who is the general counsel? Let me look in the HEW directory. You ask simple questions, but they are not that easy to answer. Have you been in Washington long?"

In a second phone call, I asked Ms. Foley what she had found out about the reports.

"You mean the industrial surveys done in the Anaconda plant. Okay, well, I'll have to call over there and find out. We don't have that. If I can get somebody in Montana who knows anything about it, you know, I'll do that. Because I can't spend too much time on this. I'm only one information person and there are information things to be done, okay?"

"Well, this is an information thing, isn't it?"

"Yes, it is, but do you think that everything else should come to a screeching halt?"

"Well, what I've requested is whatever you do have in your files."

"Well, I don't have anything but notes here."

"Whatever correspondence or notes that you can show me that you have on the Metcalf letter or anything concerning that particular problem in Anaconda."

"The letter from the union, which is the only thing I do have."

"Can you send me the letter from the union and also call me back on these two reports?"

"The '65?"

"The '65 and the '69 reports."

"And '69. Okay."

That was the last I heard from her. Senator Lee Metcalf of Montana, with more persistence, went directly to Secretary Finch for the information a month later. The bureau still maintained it did not have access to the reports. When the Senator went to the Montana State health department for the information, state officials claimed that the reports were secret under a state law designed to protect the confidentiality of the patient-physician relationship.

Only five months later, after the state attorney general ruled that the reports were not covered under the confidentiality law, did the Senator receive the reports.

The reports, most of them labeled "confidential," were shocking. In addition to the reports of alarming traces of toxic metals found in samples of workers' blood, urine, and hair, there were repeated surveys in which state and federal officials had measured fatal concentrations of cadmium fumes, carbon monoxide, and sulfur dioxide, sometimes at concentrations higher than their instruments could register.

Such findings were consistently high from year to year because the Montana occupational health program had long been weak. Director Benjamin Wake, while winning state-wide admiration for a tough stance on air pollution, had at the same time worked to undermine the occupational health program, and his attitude toward offending companies in that area was, as a Metcalf aide described it, "a little timid."

In March 1969, for example, Wake wrote the manager of the American Smelting and Refining Company in East Helena that one of his investigators found concentrations of carbon monoxide of 300 parts per million, 1,000 ppm, and 500 ppm in various areas of the plant. (Even the level for safety which industry used at that time—50 ppm—was considered harmful by some scientists.)

"As you know," wrote Wake, "these concentrations are extremely high and suggest that a fatal situation could easily develop. An immediate control program is indicated." Wake concluded: "We would appreciate your advising us what steps have been or will be taken to correct this problem."

Two months later, his remarkable faith in Montana industry unshaken, Wake wrote the plant manager again, this time to tell him that state investigators had measured cadmium levels in the plant at thirty times the industry-accepted safe level. "We felt that you would like to know these concentrations," Wake explained, "since fatalities from cadmium poisoning are reported from time to time. We appreciate your interest in this matter."

Wake could have ended such problems any time under a 1967 state law which provided power to close down dangerously hazardous operations and for fines of up to one thousand dollars a day for violating state standards.

Dick Sirginson, the state's only industrial hygienist, told me Wake worked to undermine his own program, systematically denying requests to repair broken health engineering equipment or buy new equipment, meanwhile buying desks, file cabinets, cameras, and tape recorders. At one point, Sirginson said, "Wake ordered me to find a desk and chair in an abandoned telephone switch room . . . in the basement, and move down there to make room for a new air pollution control engineer, and leave the furniture which had been charged to the industrial hygiene budget."

Wake further instructed the health and safety inspector "not to talk to the union men at all." He said Wake told him repeatedly "not to visit Anaconda smelter properties" but instead to give all his attention "to garages and dry cleaners, etc."

When I interviewed Mr. Wake, he assured me that he had everything well under control. "I think we've done a good job in Montana," he said. "I don't think anyone's fought harder for improvement than I have." After I mentioned the shocking reports that I had read of conditions at Anaconda and AS&R plants throughout the state he conceded that the smelters have "some problems.

"We have to watch them very carefully, but there has been some good work done. I can't say that they don't care." Really, he concluded, with occupational health it gets to be a problem of "how far we want to go—whether a society free of stress is desirable. To design an environment completely free of any risk seems to me would be an environment that would be uninteresting —a sterile environment."

This would have been good news to Montana workers who might have been hearing some of the popular discussion of boredom on the job. One thing Montana smelters would always have, as long as Wake had his way, was an "interesting" environment. They could also have been entertained by those interesting symptoms that so often go along with an interesting work environment—symptoms like those of Donald Olsen.

Donald Olsen was twenty-seven years old when I interviewed him in the summer of 1971. He had worked three years in the loading shed at Anaconda's East Helena lead refinery.

"One day I was feeling kind of pooped and tired and decided I'd get a blood test," he said, "and it came back leaded." Lead has been recognized as poisonous for two thousand years. Modern medical reports generally speak of lead as primarily a threat to children who eat lead-based paint, and perhaps, eventually, to all of us, if the cumulative effect of low-level exposure to auto emissions proves hazardous. Industrial lead exposure is generally considered a problem of the past. Or, if it exists at all, as a problem of "small-scale industries." Donald Olsen was proof, however, that the age-old poison was still a problem in large-scale industry. He said he was "tired and run-down a lot." He had suffered stomach cramps, nausea, constipation, and weight loss, as well as other, less easily definable symptoms. "I run around most of the time feeling like I've been beat up by a bunch of hoodlums," he said. "My wife says I'm irritable. She says, 'To put it bluntly, you're hard to live with!' " These are frightening symptoms for a twenty-seven-year-old man. "If I start having muscle spasms and go freaky, you're not going to have the Anaconda Company knocking on my door saying we'll take over for you," he said. "A woman in this world, if they're not highly educated, they can't make enough to support a family. To the company I'm not Donald Olsen with a wife and kids. I'm just a walking piece of machine called 45122.

"Sure, I'm expendable. But there's only so many of us in the whole U.S.A. So after they go through all of us then what?" Olsen's view was fairly typical of Anaconda workers. "All the jobs are the same," he says. "Everyone's got a beef about how rotten it is." Yet he continued to work because, as he explained, what choice did he have. "I was complaining to the superintendent about this multimillion-dollar cheap outfit. His reply to me was, 'You know, there are other places of employment.' I said, 'You're right, but they're all the same.'"

Anaconda workers, members of a Steelworkers local, were on strike when I visited the plant. Three men on picket duty sat around a makeshift table in the strike house at the plant gate, playing poker. The "strike house" was a shed, white on the outside, unpainted plywood on the inside, the walls spotted with thumbtacks and bits of paper where girlie photos had been tacked up during the last strike. Vandals had since broken in through the windows and torn off the photos. The two windows were still boarded up.

The three strikers represented three generations—Herbert Rate, sixty-one, a friendly, white-haired man; James Gleich, thirty-five, whose three children crowded around him on the bench; and a youth, perhaps twenty, who was working summers at the plant. All agreed that working conditions were rotten. "It gets so bad down there at the bag house," said Rate, "you can't even open your eyes. Those guys—I don't see how they can work down there. And the furnace men get that zinc dust in their hair and it's just like your hair is full of glue.

"I used to drive a couple of horses with that lead," he said. "They'll just sweat and then they'll heave, like the worst type of asthma there is.

"Working conditions are lousy as far as I'm concerned. I've seen characters there at AS&R,* their fingers just sort of formed

* The American Smelting and Refining Company had a refinery adjacent to the Anaconda plant and handled the first stages of the refining process, then shuttled the slag next door to Anaconda. Since my visit to East Helena, AS&R has taken over both operations.

like claws from the lead, kind of an arthritic condition. Those that were leaded had a blue line right across their teeth. I know my brother's had that, though I can't say I've seen that in recent years."

Why, I asked the men, do they stay at Anaconda?

Gleich answered: "I've got ten years which I'm not going to throw away. Wages are pretty good for around here and fringe benefits are good." He added, "I haven't had any trouble, but in different departments, especially the bag house, there've been several people that have been 'leaded.' This company is awful cheap. They don't want to do anything unless they make a profit. I hope to God my kids don't have to work here."

Rate agreed. "It's just the idea that a person's got to work some-place for a living. There's very little industry around here. I worked in a butcher shop for a while and in a creamery. I had an asthmatic condition from the steam; then I got over here and it kinda cleared up. There's very little industry around here—anything looks good when you're really not getting anything."

Jack Harris, secretary-treasurer of the amalgamated Steel-workers local which represented workers at both AS&R and Ana-conda plants in East Helena, 135 workers at AS&R and another seventy at Anaconda, described extensive lead exposure and addi-tional problems from sulfur dioxide, carbon monoxide, zinc, ar-senic, and cadmium. He said there "have been cases where men passed out from carbon monoxide."

Industrial hygienist Dick Sirginson said on one occasion when he visited the AS&R smelter to check on a union complaint of excessive carbon monoxide exposure, he measured 1,500 parts per million of the gas—a lethal concentration—above the furnace where men sometimes had to work. The furnace was charged from above, and the carbon monoxide rose from the charging hole. He said that when workers had to enter the area "the carbon mon-oxide was so bad it would knock one or two of them out. They kept one man stationed outside and when one dropped the man would go in and drag him out."

"I think there are a lot of people working who work day after day, particularly at AS&R," Jack Harris said, "who shouldn't be, because of excessive lead levels. We have an inordinate amount of what doctors here diagnosed as gout. I've always been at a loss to understand why. They may drink too much beer—I don't know."

Gout-like symptoms are typical of chronic lead poisoning, and usually accompany a disease of the kidneys medically termed nephritis, involving scarring and shrinking of kidney tissue. Cadmium poisoning also produces symptoms which resemble those of gout—pains in the affected bones and joints. Cadmium fumes can also produce respiratory symptoms, chest pains, inflammation of the lungs and difficulty in breathing, nausea and vomiting.

Montana state law required employers to file monthly absence reports. Purloined copies of the reports submitted by the East Helena plants of AS&R showed an inordinate number of workers reporting sick with "colds," often for eighty, ninety, or more days. Dick Sirginson said exposure to sulfur dioxide causes cold-like symptoms. Sulfur dioxide when inhaled, he explained, combines with water to form sulfurous acid, which burns the lung tissue. This of course weakens the lungs and makes the worker more susceptible to infection.

American Smelting and Refining officials were almost derisive about workers' fears and complaints. When I arrived at AS&R's management offices in East Helena for a tour, smelter manager S. M. Lane greeted me with a smile and, gesturing toward the surrounding industrial complex, joked, "We're all in the final stages, I'm sure." On the walls of his frugally furnished office were framed photos of the plant, smokestacks billowing heavy clouds of wastes—the sort of photos that were once popular among industrialists as symbols of progress but that most would be embarrassed to exhibit today.

"We have a pretty good safety program already," Lane said. "Not but what things can't be improved. They always can. Our problem is, lead is a poison. You have to ventilate as best you can —of course the workers wear respirators. I can't quite see some of these operations being that well ventilated."

The East Helena plant is primarily a lead smelter. It was built in 1896. Much of the machinery looked as if it had been around at least that long. The exterior shell is of brick, and cracks from a 1935 earthquake were visible in the brick walls of the bag house. The bag house is the final step of the refining process; it begins with a crude lead ore. The ore travels up a conveyor belt to a crusher, empties onto a second belt which carries it to a screen that allows only the finest particles to fall through. The coarse ore is crushed again. After yet a third crushing, the ore, now a fine gravel, is dumped into ore cars and carried to another building. The crushing operation was unbelievably noisy. Standing next to the crusher, trying to communicate with my guide, I found I could not hear myself shout.

We waded through an ankle-deep layer of a fine, powdery, gray-white dust in the yards. The dust clung like flour to my boots and looked to me suspiciously like lead dust. My guide seemed to be denying that it was lead, but he also said it was harmless because it is so heavy it doesn't rise into the air. Hardly reassuring. Later while we inspected another process in a room jammed with machinery and piping, the air filled suddenly with a yellow haze. We were wearing respirators, and hoping it would protect me, I took in a short breath. It didn't. The gas unmistakably was sulfur dioxide. Choking and eyes blurred, we ran for an exit, under machinery, around a corner, and up a steel ladder.

The rest of the tour was comparatively uneventful. Company officials were apologetic that I had been exposed to one of those rare "start-up problems." Later, corporate representatives arrived and told me earnestly of the company's strict occupational and environmental health program. Michael Varner, from AS&R's department of environmental sciences, did most of the talking. He was quite proud of the department, let me know that Dr. Moyer Thomas, once head of the department, invented the first SO_2 analyzer and that Philip Drinker, who is generally considered the father of industrial hygiene and is the inventor of the iron lung as well, at one time headed the department.

AS&R's problems were not nearly so severe as government agencies claimed, he said. He called the Environmental Protection Agency reports of lead, cadmium, and arsenic pollution in East Helena "biased" and much of it "completely irrelevant."

"Scare tactics is what it is," he said. "We do a complete urine check and blood analysis on every one here every six months, and some more often."

"There may be some sickness—maybe cold or the flu," added Richard Doggett, a plant safety man, "but it's no more than would be anywhere else."

In light of the repeated severe hazards cited in recent state surveys, and AS&R's history of problems since 1902, I was doubtful. So was Dr. Mary Amdur, a Harvard researcher with a long history of skirmishes with the company. I interviewed her much later at her office in the Harvard School of Public Health building in Boston. Her face wore an appearance of permanent skepticism, and after hearing her story, I could understand why.

Dr. Amdur, a biochemist, was hired in 1949 by the same Philip Drinker, then head of the Harvard School of Public Health, to work on the toxicology of sulfuric acid. Earlier that year, in Donora, Pennsylvania, a heavily polluted industrial town, a sudden temperature inversion trapped industrial fumes and smoke in the valley and within a few days caused the death of twenty townspeople and left several hundred others seriously ill. "The AS&R zinc plant was one of the industries in Donora," said Dr. Amdur. "The question was raised whether sulfuric acid was one of the things involved. Mind you, AS&R didn't give a damn whether anyone had been injured, as long as it couldn't be proved to be caused by them. They would like to go back to the Gay Nineties—when one of your workers died, you buried him and hired another.

"The funds for me to work on sulfuric acid came from AS&R. It's not at all unusual that an industry has contracts with a university. In Phil Drinker's case, how he got tied in with the smelting company, I do not know. But as far as AS&R was concerned, he was head of their industrial hygiene department. He didn't make any secret of his association. Phil was in many ways startlingly naive. Life would have been simpler for him if he had been either good or bad, but Phil wavered. He was on the fence.

"What I did not know but found out later was the fact that on this research he had said to AS&R, why don't you do it in your labs at Salt Lake City? They said no, if we do it it will be suspect, and if you do it at Harvard, it won't be, and you can keep your eye on it and make sure nothing harmful comes out."

The trouble began when Dr. Amdur began to present her findings. To her first paper, AS&R insisted that she add a statement that sulfuric acid concentrations were usually no more than .25 milligrams around a smelting plant. "I had no objections," said Dr. Amdur, "because I didn't understand the significance." She presented her second paper, which was on sulfur dioxide (SO_2), to the *Lancet,* a highly esteemed British medical journal. It was accepted, but before it could be published the smelting company howled. "Sulfuric acid was bad enough," said Dr. Amdur, "but SO_2—that was their major pollutant. They didn't like it the worst sort of way. They telephoned me. I brushed them off." Then the company complained to Drinker, who was in Belfast on a sabbatical leave. Drinker wrote to Dr. Amdur that he had withdrawn the paper "because the smelting company had ordered him to do so."

"Well, I was mad as hell. To me, within the framework of the university this could not happen. I went to the dean of the medical school. I sort of handed him this letter from Phil. He read the first three paragraphs. He said, 'Shut the door. Something stinks in here.'

"Drinker meanwhile had come back. At this point, Harriet Hardy entered into the picture. She didn't like him. She regarded him as the forces of evil. When this happened, I wrote Harriet and told her. She told me the things that were the key to being able to outwit Phil Drinker. That was that Phil had conflicts, but in the ultimate, the most important thing to Phil was being Phil Drinker, Harvard professor. What she had said—she got this from Alice Hamilton—all the Drinkers got this, breakfast, lunch, and dinner: 'At all costs you must be successful. You must succeed, you must be somebody.'

"So I went to him and I said, 'Now look, P.D., you're Phil Drinker, professor of the Harvard School of Public Health, and you can do anything you want and get away with it.' And he did, and we resubmitted."

Next she studied the effects on guinea pigs of sulfuric acid and sulfur dioxide combined. "Phil was wanting a paper to present at a meeting in May. I said I would do it if there was no malarky with the smelting company. So that was our bargain. Then the time for the meeting came near. Drinker came to me and said he was going to a meeting of the board of directors and would like to take copies of the papers along. I decided to give it to him and see what happened. I left for Chicago. A call came for me at home from Drinker. He was sort of suggesting I not go to the meeting. I had already gone. When I arrived at the hotel, Sherman Pinto, AS&R's medical director, and Ken Nelson, industrial hygiene, materialized one on either side of me. It was plain they'd been covering both entrances to the registration. They took me to lunch and suggested I withdraw my paper—not present it at the meeting. I wanted no part of it. They wrung their hands and said it would ruin the smelting company.

"I just said no, I will not withdraw it. They said Phil Drinker will be here tomorrow morning. Maybe he can get you to see our point. After this they were just sticking with me like burrs. The next morning, Phil had arrived and arranged that we have breakfast. At breakfast they gave me the same business. I reminded him of the bargain we had made. He said that was right but he hadn't foreseen some things. He suggested that I delay the paper. Give them a chance to clean up."

Dr. Amdur stood firm. She gave the paper. "When I arrived back at Harvard, Drinker told me there would be no funds available for Mary after July first. This was in 1954. I'd been there since 1949."

Within twenty-four hours, she was hired by another department. "Twenty years later," she said, "Ken Nelson said he had tried to point out to the smelting company that they were cutting their own throats. They just didn't pay any attention.

"Strange things motivate people in this field," she said. "The morality would shock the naive."

In 1969 the biggest news in Hungry Horse, Montana, was dirty air. Anaconda Aluminum, the major industry in Hungry Horse, was the culprit, and the citizens were concerned about fluoride emissions which were killing vegetation on the side of the mountain behind the plant and in nearby Glacier National Park as well. Benjamin Wake was quoted as saying, "The pollution problem in Columbia Falls is atrocious. Something needs to be done. Everyone agrees. The question is how. Every aluminum plant has the problem."

Dr. C. C. Gordon, a professor of botany at the University of Montana who was studying the problem, said the fluorides were apparently affecting wildlife as well as vegetation. He examined four wild deer captured in the area and found them "really loaded" with hydrogen fluoride. This was of potential concern, he said, because hydrogen-fluoride poisoning causes arthritis and pleuritis-like symptoms in man.

Conditions within the plant were similar and, in fact, far more serious, but since state and federal surveys were kept confidential, neither the public nor Anaconda workers could know that. Probably the most serious hazard to the aluminum workers was exposure to coal tar fumes—fumes probably similar to those inhaled by coke oven workers, although no one had yet bothered to study aluminum workers to find out if they too suffered high lung and scrotum cancer rates. A 1958 state study found "substantial exposure" to the coal tar fumes. "Ten men are subjected to massive concentrations of this smoke, ten more to somewhat less concentrates, and at least 100 to apparently significant levels . . . [and] it would appear to be of considerable concern."

Whatever the state's official concern may have been, it didn't go so far as to pressure Anaconda into cleaning up. A 1963 federal study measured "substantial quantities of tar in the total particulate which, in turn, contains benzo(a)pyrene, a known carcinogenic agent." Again, in 1968, a government observer wrote of the coal tar hazard: "At present there is no treatment of gases which arise [from the pots] . . . At the time of the viewing, the smoke within the building housing the potline was so dense as to obscure the end of the room."

During my tour of the plant it was obvious that the condition had not changed since 1953, except probably to increase, since by 1968 the company had expanded its operations more than three hundred percent. The day of the tour was hot, and as I drove to the plant I wished that I had not decided to wear the long-sleeved shirt which had been comfortable in the mountains at Glacier, where I was camping. At the plant I was introduced to Jack Canavan, Anaconda's public relations man, who was to give me the tour. I asked Canavan if we could take the union president, Chuck Forman, along with us. I knew Forman was working in the plant that day. Canavan, startled, agreed, and thanks to Forman the tour was most enlightening.

Aluminum is made in an electrolytic process in cells called pots. The pot is made basically of two parts. The lower part, the cathode, holds aluminum oxide, which dissolves in cryolite, a fluoride salt. An electric current is passed through the pot and decomposes the alumina. The top part of the pot, the anode, is made of carbon. In the process, aluminum is formed—but also carbon monoxide, carbon dioxide, and hydrogen fluoride gases. These, as well as hydrocarbons from the coal tar pitch in the anode, are a potential hazard to workers.

The pots are in rows a thousand feet long and steam and hiss, sounding like the geysers at Yellowstone. They have a heavy sweet smell, probably because of hydrocarbon fumes which fill the room with a haze. As we rode between the rows of pots in an electric cart, I was suddenly glad after all that I'd worn my long-sleeved shirt: The temperature was 140 degrees and radiated from the pots so fiercely that my exposed hands and face turned red. As we watched a siphon truck tapping aluminum from one of the pots, uncontrolled fumes poured from the top of the truck and the pot.

Forman began telling me about the dangers of "pitch blows," and Canavan explained, "We do have a problem with what they call a paste explosion when they pull the pins. The liquid pitch runs down into the electrolyte and causes sometimes a fairly violent explosion."

"I saw the whole end of Room Three on fire yesterday from a pitch blow," said Forman. "That's one job you wouldn't get me to do is pull pins. They're getting more out of them than they were ever designed for, and that's why we're having problems."

"Why do you say we're doing that?" asked Canavan.

"Well, to get more production."

"Oh."

Our next stop was at the foot of the smokestacks, and Canavan began to explain how the system controlled 98.5 percent of fluoride and hydrocarbon emissions.

Forman broke in to point out that the height of the stacks had been cut almost in half, so that the fumes would be less visible to the public. "Now I get complaints from the men that it's coming down into the plant. You can feel the mist now coming down."

Canavan quickly amended Forman's explanation. Under the old system they had produced seventy-five hundred pounds of pollution daily, he said, but with the new one were down to twenty-five hundred, two thousand of which "is going out the roof." As for the lower stacks and the additional exposure to workers, Canavan said Anaconda was in the process of doing something about that.

Forman directed Canavan to drive us to an area where ventilation fans in the ceiling had broken down. "You see those holes in the roof? That's where fans have been taken out. When they break down, they just take them out." I looked at Canavan.

"This is a top-priority project," he said.

Next we saw men cleaning out a broken-down pot. The process releases strong ammonia fumes, which rise from the pot uncontrolled. Forman said it was a common problem. "I've become so sick from it I had to go outside."

"Uh, this is a priority project," said Canavan. "We're in the process of designing to prevent that."

Back in Canavan's office afterward, without Forman there to contradict him, the P.R. man spoke more confidently. "With the new federal regulations coming in, let's face it," he said, "the states will have to comply with the standards. Of course we never really had any indication that we were doing anything out of the ordinary. The employee is the number-one priority. There is a divergence of opinion on what is a safe condition and what isn't."

Paul McMaster, Anaconda manager of industrial relations, stopped by to aid Canavan. "Anytime an employee is gone from here more than three days he is put on sick leave. We don't know of any abnormal amount of respiratory problems. If there were, it would show up."

It has shown up. "We do get respiratory problems from the plant," says Dr. David Kauffman, a local physician. "I have two

I'm treating now. One has asthma. The minute he gets in there he starts having problems. Another I've had to pull out of work. He works about two weeks and just can't tolerate it anymore. Some of these people are far more sensitive to these chemicals that are floating around. Some people come down with a cold. As long as they're working in these fumes it's impossible to get rid of. I'm sure there's months when I see ten to fifteen a month. I'd say this month I haven't seen three people. It may average about twelve a month, but I'd guess more like eight or ten. I'm sure there are some effects," he added reluctantly. "Though, I think if we could get statistics—a life span study—I don't think we'd find a great deal of difference."

To those who are affected, like Myron Jurkowski, it does make a great deal of difference.

Jurkowski has worked several jobs at Anaconda—hauling alumina ore and silicon and pulling the pins. "There's dust from the truck and you're above the pot and above the gases," he said. "I'd been getting sick, off and on, for over a year. I went to Dr. Kauffman. He said I had a bad strep throat and should lay off for about six weeks. I work until it gets sore and then I have to quit. I feel a lot better after being out of there."

He was also exposed to chloride gas from the casting building. "Sometimes it backfires on a windy day. You get dust from the floor; then that chloride from the furnace backs up. That's what really gets you. One night I was working graveyard shift, and I just got sick. I was vomiting. I thought it was from chewing tobacco. Everyone out there chews gum or tobacco—it picks up the dust in your mouth. After chewing gum for a while, it gets gritty.

"Working out there, I'd get so dopey, like if you'd want to go out and do something different, like chop wood or something, you just couldn't. It just draws everything out of your body—the sweat and the heat, you know.

"It's just the opposite in the wintertime. When it gets below zero, it's right comfortable."

The gases have affected a worker named Donald Amundson so badly that he had to quit work at the Anaconda plant. He said he had gotten "a touch of emphysema" from exposure to the fumes and chemicals. Though he had suffered from asthma since he was a child, the twenty-six-year-old worker said he never had any lung problems until he began working at Anaconda. His first real problems came when he got caught in chlorine gas.

"I got this dizziness, a feeling you want to vomit. It immediately clogged up my lungs. It took me a good hour to sit outside and recover from it."

Eventually he had to get out of the plant. He stayed out seven months, then went back. "I started getting my chest problems again. Pains across the chest. Hard to breathe. The same symptoms I had before. A slow death is probably what you want to call it."

Amundson said most of the workers did not like to admit to themselves that the fumes are harmful, which is understandable, since they have to continue to work in them.

"Anaconda employs a thousand people out there," said Amundson. "If they close down, that's a thousand people out of work."

CHEMICALS

"We're the Guinea Pigs"

If you have emphysema or other chronic lung problems, you know what it's like to climb a flight of stairs. And you probably don't know what it's like to play a round of golf or ride a bike or even take a walk anymore.

Union Carbide's Linde Division has developed a portable liquid oxygen system which many doctors are prescribing for their patients.

Sling the handy carrying case over your shoulder. It weighs less than 9 pounds full. Set the oxygen at the flow your doctor tells you to. And you can do many of the things you did before.

Sure we've oversimplified the whole thing. We're not going to go on and on and on about all the Union Carbide technology that makes the Oxygen Walker possible. It's

just one of the things our Linde Division is doing with
air . . .

    —from a Union Carbide advertisement in *Business Week,* 1970.

Linde Division *has* done lots of things with air. It has produced the
Oxygen Walker for emphysema patients, and in Tonawanda, New
York, it has produced patients for the Oxygen Walker. In 1970,
union officials of the Oil, Chemical and Atomic Workers documented seven cases of emphysema at the company's Tonawanda
plant in a department employing eighteen workers. Other workers,
the union charged, suffered bronchitis, dermatitis, and leg ulcera-
tions. Not all seven men suffering emphysema worked in the department at the same time, and by 1970, none were there—it's
hard to work when you have emphysema—but still the figures were
dramatic.

When I visited Harold Michel, one of the affected workers, I
arrived unannounced because he had an unlisted phone number.
I introduced myself, and told him why I'd come. He smiled
broadly and said, "You're just the girl I've been waiting for. Come
in. What can I do to help you?" A short, wiry man, he showed me
into the living room, left and returned with four bottles of powder.
He opened them and had me smell each one. "The reason I got
emphysema is because I worked in powders like this," he said.

Michel had worked in Linde's molecular sieve department,
which produced a gray powder used in refrigerators, car exhaust
systems, Thermopane glass and oil refineries to absorb moisture.
Michel took the bottle containing that powder and poured a little
of it in the kitchen sink so I could see how it absorbed water.

Michel had worked in shipping and spent much of his time
filling customer orders. It was a particularly dusty operation, he
said. It involved scooping the powder—and other materials, including various caustic powders and muriatic acid—out of fifty-
five-gallon barrels and into small containers for shipment.

"The dust goes right through your clothes," he said. "Your skin actually dries out." The powder, he added, went through his respirator, too. "It would get in your mouth—stick right on your teeth."

When I visited him in 1971, Michel was still working, though no longer in the molecular sieve department. After his emphysema had been diagnosed, he said, "they give me a job picking up papers outside."

Michel, at fifty-four, was by nature an active man and, in talking about his illness, energetic and excitable, often repeating himself in his haste to tell his story to someone who he hoped could help. His wife, Ruth, listened silently as he talked. She made no comment during the forty-five-minute conversation in the house or later during lunch at a nearby diner. Only as we were driving back to the couple's house, and after Michel had finished his story, did she suddenly speak.

"If he goes another place for a job," she said in a low angry voice, "they won't hire him. If he ever dies," she said, staring straight ahead, "and it's caused by emphysema, I'll sue the damn hell out of Linde. He didn't have that when he got out to Linde, and he got it out at Linde, and they're going to pay. I'm not going to be like these other widows that let them get away with saying it was a heart attack."

Joseph Shoemaker was another Linde worker with emphysema. Shoemaker said he didn't realize he had emphysema until he was hospitalized for a hernia and doctors diagnosed the lung condition. "When they told me I had emphysema, I knew it was from the darned dust. I thought, 'Boy oh boy, this really happened.' We have periodic physicals at Linde. I was *never* notified to see my doctor or anything else. I had a physical, X-rays taken not too long before that at Linde, and they never said anything about it. Either their machine is defective and it's not picking up what it should, or they didn't want it to!"

Shoemaker had noticed difficulty breathing much earlier but

hadn't worried about it. "Heck, it was almost a natural feeling in that place," he said. "The dust was so heavy it would accumulate on top of pipes like a pyramid. Part of my job was changing light bulbs. I'd shake the conduit to change the light. Boy, it would come down like snow. The loading hoppers were bad too. Sometimes we would have to go in there and work and it was terrible dusty.

"These were conditions you accepted in taking the job. I made comments when I was working there that the ventilation system was inadequate, but I figure it's my fault for not putting more pressure on the people than I did."

I had hoped to talk to more of the men whose emphysema the union had documented, but by 1971 two of the seven were hospitalized—or so their fellow workers believed, though they did not know for sure where—and three had died.

In August 1970, Don Kreuter, financial secretary of the OCAW local at Linde, called an official of the international union to report that the company was laying off one hundred and one employees and putting about a quarter of those on early retirement. "Most of these people are welders," Kreuter reported to the union official, "and the basis for the decision on retirement are indications of emphysema and heart problems."

Chris Steurenagel had worked at Linde as a welder until 1969, when his doctor told him he had emphysema caused by breathing welding fumes, or what workers call "welder's lung."

"He had looked forward so much to the time when he could retire and play golf," Mrs. Steurenagel told me. "Now he can't breathe. We're five houses from the corner. He used to walk to the corner to buy a paper. But he can't do that anymore, because he'd have to stop halfway back and bend over or squat until he could breathe again."

Steurenagel was collecting $191.20 a month in Social Security disability benefits. He had applied for workmen's compensation as well, but the hearings had been postponed time and again and his wife said they had just about lost hope.

I called John Patti, Steurenagel's attorney, to find out more. At first Patti had trouble recalling the case and said he believed it had been favorably disposed of, but eventually he remembered it had not. He explained that Steurenagel's hearings had been postponed because of "problems getting doctors in for testimony."

I talked to Mrs. Steurenagel again and mentioned my conversation with Patti. "We're just another number on his book," she said. "Now there are six or seven or eight doctors involved in this, and you're never going to get them all together at one time." She said that at one point her husband received a notice that a hearing had been postponed—and he had never been notified of the hearing in the first place.

Mrs. Steurenagel sighed. "Of course, the first shock is worn off since the day when the doctor said he couldn't go back to work again. We keep thinking—all those years he worked there. He paid into the fund—disability benefits, retirement fund, all these funds. He never missed a day for twenty-two years. We thought there'd be something coming. The personnel director at Linde did say if we'd say it was non-occupational they'd send something. I feel they're against us, and they're too big. We can't win."

In August 1973, I read in the Oil, Chemical and Atomic *Union News,* the OCAW's official newspaper, that Steurenagel had finally won his case and would receive thirty dollars a week in compensation. "The Compensation Board had originally ruled in Steurenagel's favor, but the decision was challenged by the State Insurance Fund of New York," the *News* story said. "Then his original lawyer quit the case, telling him that he would receive nothing from the Board. At that point, Donald Kreuter, financial secretary of the local, contacted another labor lawyer, Richard Lipsitz of Buffalo, who took the case and pursued it to a successful conclusion."

In 1971, Union Carbide spokesman Dennis Holt denied all

charges of workers contracting emphysema from plant exposure. Holt's argument was difficult to follow. "There is no medical evidence that these guys have emphysema," he said, and then somewhat contradicted himself: "We're not denying that these people got these illnesses. But it's impossible to prove it came from the job. We question the irresponsible allegation that he worked at Union Carbide, he had emphysema, therefore he got it at work. It's rather unfair."

It is difficult to prove the cause of such a disease, thanks to the retarded state of occupational disease research in this country. Dr. Harry Sultz, a University of Buffalo dentist who had made several studies of Buffalo air pollution, became interested in the problems at Linde. With the union's cooperation, he began a study of the molecular sieve department workers in 1971, comparing their chest X-ray findings and lung function tests to those of workers in other departments. If the molecular sieve department showed a higher rate of abnormalities, he reasoned, they'd be able to prove a causal relationship. But early on in the study he began to doubt that his findings would be conclusive, not because of a lack of lung pathology among the molecular sieve workers, but because of an unexpectedly high incidence among the general plant population as well. The study was incomplete at this writing.

Union Carbide refused to give me a tour of the Tonawanda plant for "proprietary reasons," though I suspected the refusal had more to do with the condition of the plant. "The building is in bad shape," Richard Elsaesser, a union steward, told me. "The roof is in bad shape, windows are broken, doors are off. When silicones were running, we had a real good safety program, fire drills, all kinds of safety meetings. Silicones were showing a profit. We had good supervision, we had people interested in what was going on. Now we don't have good strong supervision. But there's no reason for it to be that way. You can make a profit and still be safe, I say."

The Oil, Chemical and Atomic Workers Union has been a leader among labor unions in the area of occupational health, but often even the most enlightened union leadership is powerless to protect its workers from early death, because no one knows what the results will be of exposure to any of thousands of chemicals. The number of chemicals is multiplying rapidly. The 1958 edition of *Chemical Sources,* a handbook that describes itself as "designed to keep the chemical buyer up-to-date," listed 17,000 industrial chemicals. The 1970 edition listed 41,000. Other sources estimate as many as 100,000 to 500,000 chemicals are in use in industry. No one knows what most of these new chemicals do to human beings. Indeed, because research funds are limited and the effort is slight, American scientists know little about even the most common chemical working hazards. Indeed, scientists are still actively debating the "safe levels" for such long-recognized hazards as lead, cotton dust, and asbestos.

In a recent study of an Anaconda refinery in Montana, investigators found traces of cadmium, lead, zinc, copper, and arsenic in workers' blood samples and high levels in urine and hair. Yet they concluded: "Lead is probably the only element which is currently of any significant hazard to the workers." The state health department had reached its conclusions by default. As the report explained: "A clear relationship between levels of elements in blood, urine, and hair with possible toxicological effects has not been established." Imagine the public outcry if the same logic were applied to food and drugs. Cadmium, for example, is considered too toxic to be permitted in *any* concentration in a food product. And arsenic, as Rachel Carson pointed out in *Silent Spring,* is "the environmental substance most clearly established as causing cancer in man."

In 1974, government standards were available for only some 450 hazardous substances. Most of those standards, including one for lead, did not go into effect until August 1971. Earlier that year, when Labor Department officials inspected an American Smelting and Refining Company plant in Omaha, Nebraska, they found lead exposure above the generally accepted level of 0.2 milligrams per cubic meter of air over an eight-hour day. But since standards had not yet been set under the new Occupational Safety and Health Act, Labor Department officials could only cite the company under a "general duty" provision of the act, which requires an employer to provide a work place free of "generally recognized hazards." And the company appealed the citation, arguing that lead was not a generally recognized hazard! (The case was on appeal at this writing.) That kind of corporate arrogance leaves one with little hope for the safety of millions of workers exposed to the thousands of chemicals whose toxic properties are yet unknown.

A chemical known as tolylene diisocyanate, or TDI, was first put to industrial use in the early 1950s. By 1960 it was widely used as a polymerizing agent in an enormous variety of polyurethane foam—or "foam rubber"—products. TDI was soon found to have the particular property, however, of causing acute asthma-like sensitivity in some of the workers exposed to it, but apparently not in all. In response to this observation and a limited number of animal studies, an ACGIH committee * set a threshold limit value of .1 parts per million—which was lowered in 1963, on the basis of new studies, to .02 ppm.

In the late 1960s, Dr. John Peters of the Harvard School of Medicine and several co-workers were studying employees in a factory which manufactured rubber dolls, and they found workers were suffering chronic lung damage from exposure to TDI at levels far below the supposedly safe .02 ppm. The effect of TDI on the lungs, he explained in an interview, is "essentially obstructive, causing acute or chronic bronchial narrowing." The researchers measured lung function with several tests, including one for "forced expiratory volume" or FEV. The average decline of FEV in a

---

* The American Conference of Governmental Industrial Hygienists—a quasi-official body which often represented industrialists' views more faithfully than those of the public—was for years the only agency regulating toxic substances in industry. ACGIH standards were used by most states and were later incorporated as standards under the Occupational Safety and Health Act of 1970.

normal person is .02 to .03 liters per year, or .12 to .18 in six years. But twenty workers examined by the Peters group experienced an average decline of .11 liters per year, and .22 liters in two years—far worse than the normal decline in lung capacity. Since an average man has a four-liter lung capacity, such a decline, if exposure continued for five years, would logically result in the loss of one-fourth of lung function. And some workers, Peters found, respond to TDI far more drastically even than that. He saw one forty-year-old man who was exposed to TDI for three weeks, claimed he was in good health before the exposure, but during the five years since has been completely disabled. "That's rare," said Peters. "But it can happen. That's what's scary to me."

Herbert Stockinger, the chairman of the ACGIH committee on threshold limit values, was apparently unimpressed, however. The 1971 edition of his committee's *Documentation of Threshold Limit Values* cited Peters' work but recommended a continuation of the .02 ppm level because, "In the opinion of the TLV Committee, the changes noted by Peters' group . . . are not of sufficient importance to invalidate this limit."

In March 1973, I talked to workers at a Minnesota Mining plant in St. Paul. They were exposed to TDI in several processes and they were worried. Men were suffering a variety of symptoms —from passing out to chronic shortness of breath—but as one worker told me, "The company said it was safe because it was always under the threshold level."

Local OCAW union officials estimated that about three hundred 3M company employees worked in areas where they were exposed to TDI. "Most people around here would rather suffer conditions than be transferred, because they would lose their department seniority," said Art Pottoff of the OCAW local. "They don't want to admit they have health problems."

Even so, I talked to fifteen workers who admitted that they were suffering, some of them apparently from TDI exposure. Some of them said they had serious lung problems. One was Tony Van

Heel, a fifty-eight-year-old worker with twenty-eight years in the plant. "Where I run into this," he told me, "was running a molded products maker—we were making bumper buttons that go on the back of chairs. This was in the spring of 1971. When I went in there I was okay, working eight hours a day, missing very little time. I was in there a couple of months when I started complaining about my breathing. After a day's work, I could barely make it out to the car. I'd sit there until I could catch my breath, and then I'd drive home. I started missing a lot of time. I'd get shortness of breath and couldn't breathe. I went to see my doctor. He thought I just had the flu for a while. Then he said I had emphysema. I transferred to material handling, but I left after four weeks in there. I was driving a fork lift, but I couldn't even take that anymore. My lungs were shot."

When Van Heel left work in January 1972, his lung capacity was half that of a normal man. He said the Minnesota Mining doctor, A. E. Sethre, had told him his trouble was "nerves."

"They didn't do anything till you pretty near died," his wife recalled, "and I had to come get you at nineteen degrees below zero and take you to the hospital."

"I had to go back and forth, inside and outside," explained Van Heel, "and after a while I went to the nurse, and rested an hour or so. I got up and started to drive home. I couldn't breathe and I hollered for help. They called my wife, and she came in a cab and picked me up."

"It got to the point," Mrs. Van Heel said, "where I was scared when you went to work that you'd try to drive home and couldn't make it."

"I used to love to hunt," he said. "I can't go hunting anymore. And bowling—I used to be a 160 bowler. Tell her what I did the other day—104. I used to like to have a bottle of beer during the day. I can't do that. And of course, cigarettes, that's out. I can carry in five bags of groceries, and then I'm done. I suffer terrible from gas pains."

"You spend a fortune for Alka-Seltzer," his wife said. "And that stinking disability—that's starving-man's wages. That's what that is."

Ralph Sward, thirty-four years old, another Minnesota Mining worker, blamed his lung problems on TDI. "When I'm exposed to that stuff, my lungs fill up," he said. "I start coughing. That leads to vomiting. And that leads to going home." In less than two years, Sward has had pneumonia four times. He has lost one-fourth of his lung capacity.

John Schowalter first worked with TDI in 1969. "I worked in Department 56 for two weeks and it didn't bother me—at least I didn't think it did. Then I got pneumonia. No one knew it—not even I. It was hard to breathe. I couldn't walk far. I would be off a few days, feel fine, be back four hours, and I'd be off again."

"One time I heard the car drive up," said his wife. "I knew he was home. And he didn't come in, and I went down and there he was, hanging on the front porch."

"When you get it, you have to lay down and relax the best you can," said Schowalter. "The minute you move, you're huffing and puffing again."

Schowalter tried to get into the pipefitters apprentice program. The head of the program told him if he was sensitive to TDI he wasn't going to get the job. Pipefitters—and all maintenance men —travel from department to department to make repairs and are necessarily and routinely exposed to TDI. The company medical department assured Schowalter he was not sensitive to TDI, however, and he began training.

"There were three of us apprentices hired," he said. "Two went to the main plant, and I went to research. After six months we had to rotate. At research they have better ventilation and not the volume of TDI like there is in Department 56 and the fifth floor of Department 20. Two weeks afterwards, in the main plant building, I had a job in 20 on the fifth floor. We were working on a Viking pump. On one of the machines the pump was leaking a fluid. I had my nose about six inches away from the chip pan and all of a sudden I started choking. The foreman said, 'Are you having trouble breathing?' And I said, 'Yeah.'

" 'Do you know what that is?'

" 'No.'

" 'That's TDI.'

"I started to the elevator. I never made it. I got it that quick. At the nurse's office she gave me oxygen. The next day I went back to see one of the doctors—Dr. Sethre. I parked far away. Several times on the way in I thought I was going to die. I'd sit and sit, and my ol' heart, you think your heart is coming right out of you, and your chest is hurting trying to breathe. Finally I made it down to the medical department. Dr. Sethre gave me a prescription for a drug for asthma. And that was like a miracle drug for a while. I came back down to Building 26 where the shop is, and the first or second day, I was eating lunch at a workbench. Some pipes had been brought in and, sure enough, I began to choke again. What was in that stuff was TDI. So I was off a few more days.

"Then I was sent back out to research. Then they wanted to get rid of me. I fought it and won it with the help of the other steamfitters. Why go after the man, go after the chemical, I say." Since his exposure, the thirty-year-old pipefitter has been short of breath.

TDI is odorless, and it hits workers who are sensitive to it without warning. One man said he began choking up once after simply being exposed to contaminated clothing. Another worker suffered asthma-like attacks at home every time he got near a new chair which was filled with foam rubber. Yet for workers who don't have sensitivity reactions, TDI may be even more dangerous, said Dr. Peters, because long-term exposure results in reduced lung function, but without any other warning symptoms, such as chest tightness or wheezing.

When I showed Peters results of my interviews with Minnesota

Mining workers, he said, "A lot of these sound essentially compatible with TDI exposure. They sound like they're sick and it doesn't surprise me." I asked him about the many workers I had talked to who had been told they had pneumonia. TDI effects, he replied, could easily be misdiagnosed as pneumonia. "It's common to have chronic racking coughs. X-ray changes have been reported and these symptoms could easily lead to a diagnosis of pneumonia."

However, in the two rounds of talks I had with Minnesota Mining officials at the company's corporate headquarters in St. Paul, they denied any problems. I met with several public relations men, including the executive director of public relations. I also met with John Pendergrass, an industrial hygienist; Dr. A. E. Sethre, assistant medical director; Marvin Adams, senior manufacturing specialist; and Howard Lane, personnel manager of the St. Paul plant.

"We have no instance of anyone ever having been seriously hurt with TDI," said Dr. Sethre, a distinguished-looking man, with an unflappable air. "Some irritation symptoms, temporary in nature; they settle down quickly. Some are sensitized, and we have them out of any TDI area. But we don't have any evidence of chronic cases. No."

I asked Sethre whether he was aware of Peters' study. He said he was. "The Peters report certainly attracts attention," he said. "We're watching for confirmation of his findings. The preponderance of opinion indicates effects are of a temporary nature. You don't come all unglued with one report."

Pendergrass, one of six corporate industrial hygienists, said TDI exposures at the St. Paul plant "will generally be at levels of .005 or less" (.02 ppm being the established "safe" level). Adams, the manufacturing specialist, brought in blueprints. "First of all," he explained, "you only work with a closed system, so there's no chance of exposure. The places where we use it, it's zero. We go back and measure it, and it's zero. In the tape-coater department we have never found a level above .005. The policy here is, you just don't take any chances."

Lane, the St. Paul personnel manager, said the union participated in monthly safety tours of the plant. "The employee at any time can point out a hazard of any sort—and it is done. He also has the normal grievance procedure. In the last year we have not had anything appear [on TDI]. The point is, they have an avenue to bring this up."

I talked to union official Art Pottoff later. "The problem could sure as hell exist without grievances existing. We have to be knowledgeable and we're not. If they want grievances, we'll write a hell of a lot of grievances."

Jim Bergeron, international representative for region 6 of the OCAW, mentioned another problem: "Many people that work here are dedicated to Minnesota Mining. They've been convinced all these years that Mining wouldn't do them any harm. We have difficulty getting them to fight the big company."

Early in 1973 a strange nervous disorder began to afflict workers at the Columbus Coated Fabrics plant in Columbus, Ohio, a division of the Borden Company. Thomas Meade, for example, a twenty-two-year-old machine operator in the print shop, noticed something funny was happening to his arms and legs. He didn't have the strength in his fingers that he usually did. He had been a good runner, but he noticed that he couldn't run as fast or as hard or as well as he used to. His symptoms were vague and unfamiliar, however, and he shrugged them off. By June he was having trouble walking. His family had a history of arthritis. Meade thought maybe that was what it was. He went to see his family doctor. But the tests for arthritis were negative, and his symptoms worsened.

Meanwhile other men in the print shop were beginning to complain of similar problems. Eventually Dr. Mary Gilchrist, a resident in neurology at Ohio State Medical School, who was treating one of the men, suspected a toxic origin for the disease. On August 22 she contacted the Ohio state health department. The state contacted company and union officials and began testing the other 950 workers for signs of the disease. At first it seemed that the problem was confined to the print department, where at least fifty-seven workers tested positive, indicating potential nerve deterioration.

Corwin Smith, local president of the Textile Workers, said later that the union had "a frustrating time, initially, in getting someone to take action. First," he said, "we asked the State Health Department to close [the print shop] down, and they refused. We asked the company, and they refused. And so we did it." On September 7 the local Textile Workers executive board members stationed themselves at the gates and advised workers in the print shop not to report to work.

When evidence of the disease began appearing in the mixing department too, the union again asked first the State Health Department, and then the company, to shut it down. Again they were refused. The union then walked out of that department too. Smith said the union also asked Peter Schmitt, area director of the Occupational Safety and Health Administration, to close the plant when other efforts failed. "He said he couldn't," Smith told me, "because he didn't have enough evidence. He promised me that OSHA would become involved and that he personally would become involved."

"Did an inspector come?" I asked Smith.

"Yes," he said. "After the hygienist working in his office got back from vacation ten days later."

The union had called on OSHA earlier that year because of frequent fires—three or four a month in the print shop—as a result of high levels of flammable vapors. "This was not a new thing," Smith told me. "For years we've complained about poor ventilation and argued about it. The company didn't want to change it.

It's costly, for one thing. The company had a pet phrase: 'It doesn't put out any production.' "

OSHA inspected the plant on April 17, citing twenty-eight violations of the law, including "failure to take adequate precautions to prevent the ignition of flammable vapors," and "failure to keep combustible waste material and residues in a building or operating area to a minimum." All of these violations the labor department deemed "non-serious." During a second inspection on June 15, the company was cited for six additional violations, five of them involving hazardous exposures to such toxic chemicals as antimony, lead, chromium, and 2-Butanone. These, also, the labor department termed "non-serious."

Though the first signs of the occupational epidemic had already appeared in the print shop workers, labor department investigators reported no health hazards involving ketones—the chemicals later implicated by health officials as causing the nerve disease. There was no reason the officials would have been particularly suspicious of the ketones, which were used extensively in the print shop as solvents, because ketones had never before been implicated in an epidemic of this kind. Nor were there reports in the scientific literature of "peripheral neuropathy" caused by ketones.

Other agencies quickly became involved. The union hired an independent consultant, as did the company, and the union asked scientists from the National Institute for Occupational Safety and Health (NIOSH) to make an evaluation of the hazard.

When it became apparent to the union that OSHA would not take action, Corwin Smith asked the Ohio Industrial Commission to intervene and close the plant. The commission sent investigators into the plant and set a date for a hearing, serving notice on Borden that the company must show cause why the plant should not be closed. Borden appealed the order to the U.S. District Court, arguing that state jurisdiction was pre-empted by the federal Occupational Safety and Health Act. The judge, ruling in the company's favor, granted an injunction against the state order. (The ruling was on appeal in December 1973.)

When the state health department finished screening Borden workers, one hundred and twenty-eight were found to have positive findings. These workers were examined more thoroughly by Dr. Norman Allen, chief of neurology at Ohio State Medical School. According to a Textile Workers Union summary of his findings, fifty-three workers were found to have "definite polyneuropathy (including five whose conditions may be unrelated to industrial exposure). Thus a total of sixty-six individuals have probable or definite diagnoses of polyneuropathy related to their industrial exposure."

Alarming as these figures were, no state or federal agency had both the authority and the conviction to shut down the plant. Workers stayed out anyway. By the first of October most of the plant was out. By November health experts had concluded "tentatively" that the likely cause of the disease was "exposure to organic solvents in general, and to a ketone or ketones in particular." This was the conclusion in a statement of recommendations signed by four health experts representing the state, NIOSH, the company, and the union. The recommendations included eliminating one ketone, methyl butyl ketone, or MBK; monitoring the air; using personal protective equipment; and ventilation maintenance and better work practices to reduce exposure to solvents. These steps, the experts agreed, would "provide a reasonably safe working environment for employees of Columbus Coated Fabrics."

Union members went back to work, but Corwin Smith, local union president, remained unconvinced that the problem was under control. "I don't think it's MBK," he said. "I never did. MBK use was very limited. Some of the people who used MBK the most weren't affected." And if MBK hadn't been the villain, then it was possible that the crippling substance was still in use and would harm more workers.

"We're uneasy about being back to work," Smith admitted,

"but we really had no choice. The agreement was signed by four medical experts and in a court hearing [if we stayed out] that wouldn't put us in a very good situation."

In this case, the local union disagreed with its own national office and experts, as well as state, federal, and company opinion. There were others who disagreed, among them a federal epidemiologist named Richard Lemen. Lemen, a thorough and conscientious researcher, explained in painstaking detail the procedures and reasoning that he and other investigators had used. MBK was one of several ketone solvents in use in the print shop, and the most recently introduced. It had been brought into the plant to replace methyl iso-butyl ketone, or MIBK, because MIBK had caused air pollution problems. Another solvent, methyl ethyl ketone, MEK, continued to be used. MBK was suspected, he said, because no problems had occurred before its introduction. The epidemiologists surveyed 160 of about 180 men in the print shop and found out, among other things, that the men were "practically bathing in the ketones." The solvents were used for cleaning ink off machines, washing hands and clothing, thinning ink, and even mopping the floor. "Very poor work practices were the biggest factor, along with poor ventilation, which was very, very inadequate as far as removing vapors of solvents," Lemen said. "I have to say that probability points more strongly to MBK, but we still don't know *for sure* whether it was MEK, or a contaminant in MBK or MEK or possibly something else. They're still using MEK. It would be my recommendation to discontinue using both MBK and MEK until we find out for sure."

Lemen said that although he knew of no cases of peripheral nerve disease from either MEK or MBK which had been previously reported in the scientific literature, he knew of isolated cases in Iowa and Connecticut of workers exposed to MBK. He was also following up on reports of other cases among workers in plants producing MEK. "I'm just not willing to draw the conclusions that it's MBK," he said.

But the doubters were overruled. The men went back to work. "We're the guinea pigs," said Corwin Smith. The state continued health monitoring of workers, but Smith said the state had refused to release results to the union ever since the men went back to work, although union officials had previously been kept informed. "I've been told they consider this as medical information and the layman has no right to it."

But why the change? I asked him. Well, Smith said, he guessed his losing his temper at the last meeting of the interested parties might have had something to do with it. He had gotten irritated during an argument over language in the joint agreement.

"The draft said that in the event of further problems the state health department would immediately investigate and immediately take action. The company objected to that. Then a spokesman from the state health department said they didn't have authority and maybe NIOSH should be the one. Bobby Craft from NIOSH said NIOSH didn't have the authority either. I blew my top and told them exactly what I thought of all the state and federal agencies. I said we would never again wait two and a half months for someone to take action.

"After my passionate outburst," he added, "an OSHA hygienist said she would immediately file for injunctive relief in court if there were further problems."

Smith had good reason to be suspicious of the experts. What had they done, after all, for Thomas Meade, the first victim of the disease? When he became disabled in August he had worked at Columbus Coated Fabrics for only two and a half years. Twenty-two years old, with a wife and a baby, Meade was the most severely affected of the workers. He could walk only with the help of leg braces and was unable to use his hands for such simple tasks as turning on a lamp. Doctors told him there was no hope that he would improve.

"If I stay the same as I am, I don't think I could do another job," he said. "Emotionally, I feel less of a man, where I can't support my family. It's really upsetting to my wife that I may never be able to function like I did before as a worker and support my family like I did before."

Most of the other workers appear to have recovered from the effects of the toxic exposure, although it is conceivable that the chemical or chemicals could later cause long-term delayed effects. No one knows.

About the same time that the Columbus epidemic was discovered, another chemical that causes nerve damage, carbon disulfide, was affecting Textile Workers Union members at an FMC Corporation rayon plant in Nitro, West Virginia. Carbon disulfide is a well-known poison, recognized as hazardous as early as 1851 in France. The effects of carbon disulfide are particularly dreadful. Its assaults on the human body include damage to adrenal glands, the intestinal tract, kidneys, heart, blood, and blood vessels, as well as injury to the nervous system. The chemical may also cause severe psychological effects, as described by Gordy and Trumper in a 1938 report: "There may be excitement and increased psychomotor activity, flight of ideas, outbreaks of irritability and unmotivated rage. On the other hand there may be somnolence, semistupor, or simply forgetfulness and retardation and intellectual impairment. Periods of amnesia, sleeplessness, sleep reversal, horrible dreams, and anxiety states may precede the onset of definite psychosis. The psychoses may be characterized by a definite manic-depressive picture, with either manic or depressive symptoms or an agitated depression. Visual and auditory hallucinations and delusions of persecution may color the picture: delirium may occur or a schizophrenic syndrome with catatonic excitement."

Carbon disulfide was once commonly used in the rubber vulcanizing process, and in a British raincoat factory, carbon disulfide poisoning was so frequent that the company reportedly installed bars in the windows to prevent workers with suicidal impulses from jumping out.

Alice Hamilton, in her book *Exploring the Dangerous Trades,* published in 1943, wrote that although the use of carbon disulphide had almost disappeared in the rubber industry with the introduction of a new process in the 1930s, new cases then began to appear in the new "viscose rayon" industry. She wrote of the effects on workers at a Pennsylvania rayon plant where the men "knew that a distressing change had come over them, one they could not control. It spoiled life for them, it ruined their homes, it broke up friendships, it antagonized foremen and fellow workers, it made day and night miserable. They knew it was the job that caused it, but neither doctors nor employer would admit that and their bitterness and anger over this injustice was great. Later publication of medical findings and the passage of a workmen's compensation law in Pennsylvania, however, "resulted in radical reform," she wrote. "It is safe to say now that no large viscose rayon works is a dangerous place to work in and probably few of the smallest ones." Dr. Hamilton was an incurable optimist, and she wrote with conviction that "no matter what dangers may be brought about by new methods of manufacture, they can be controlled and they will be . . . The medical profession will never again neglect industrial diseases, the employer will never again refuse to assume responsibility toward them."

FMC Corporation's American Viscose rayon plant, in southern West Virginia, was first an explosives plant, built by the federal government in 1917 to make gunpowder. The town which grew around the plant, along the sides of the narrow Kanawha River valley, was named Nitro, after the nitrocellulose process at the plant. In the 1930s the plant was sold to the FMC Corporation and converted to a rayon process, which also used cellulose.

The valley today is an amazing sight. Down the center is the Kanawha River, and stretched along its banks, for miles and miles, are huge chemical plants. A set of railroad tracks and a highway run parallel to the river, and in the space that remains, several rows of houses are crowded, and these communities are never out of sight or breath of the chemical factories.

Even in this valley of chemicals, the American Viscose plant was held in ill-repute, according to Jim Weeks, an occupational and environmental health consultant to the Urban Affairs Center at West Virginia State College. Weeks had been working with the Textile Workers Union to improve conditions at the plant. "It's an industrial slum," he told me. "The black sheep of the valley, the way other management looks at it."

In 1973, when I first heard of problems at the plant, I talked to J. P. Barnette, at the time business manager for the local. "When we really first got concerned about it," he said, "was when we had people having, like, nervous breakdowns."

Barnette has since been replaced by a new business manager, and part of the problem, according to Weeks and workers from the plant, was that Barnette never did get concerned about it. Weeks said the union had suffered for years from mismanagement. When workers began having symptoms consistent with carbon disulfide poisoning, Barnette brushed them off. "He didn't want to help," said one worker. "He said, 'Well, I worked down there once, I drank a lot of beer, I never had any trouble.' Our union at that time would try to relate it to drugs."

Ronald Sayre, a worker who suffered from CS_2 exposure, said of Barnette, "He wouldn't help me any. He said he'd go with me to my compensation hearing, but the day it came, he wouldn't come with me."

Sayre began work at the plant in May 1969. In August he was hospitalized with what he called "a nervous breakdown."

"The company doctors diagnosed it as acute schizophrenic reaction," said Sayre. "It really wasn't. It was carbon disulfide poisoning." Sayre had worked as a cutterman, in an area where carbon disulfide fumes are the strongest.

Rayon fiber is formed in a chemical process. First, cotton and wood pulp are soaked in a carbon disulfide bath. A honey-colored "viscose" liquid is formed, which is then forced through jet sprays into a sulfuric acid bath. When the spray hits the acid, the fiber is formed. The fiber is gathered by machine into a thick bundle, called tow, and wrapped on spools. Cuttermen tend these machines—seven men to a shift for twelve machines. Whenever the machine malfunctions, they cut out the bad tow—and are exposed to the ever-present carbon disulfide fumes, and hydrogen sulfide as well. Often the discarded tow lies in thick bundles on the floor, sending more fumes into the air and causing additional exposure.

During his first few months in the plant, Sayre's personality began to change. "The stuff had such an effect on me, I was sick and didn't realize it. I worked over there thirty days, my wife tells me, and I don't remember going to work and coming home. I couldn't sleep. It was making me think weird things. I thought everybody was watching me. I remember one night, I punched out at ten o'clock, and I thought my father was outside, in the swamp, waiting for me. I went out and looked for him. He wasn't there."

In August his family persuaded him to go to the hospital. He remained hospitalized for thirty days. Three other men from American Viscose, with similar problems, were there at the same time, Sayre said. Three months later he returned to work. But his problems continued. Sayre's wife divorced him in 1970. "It was me," he explained, "my temper, not cooperating with her. Mean to her. But I couldn't help myself. I didn't want to be that way." After the divorce, Sayre was understandably depressed. "I stayed by myself a lot. I had a house trailer, and instead of going out, I'd stay in, watch TV. I'd think about things."

His ex-wife and their two children were living just down the road. Eventually, he said, "we got to seeing one another again. We decided we'd done the wrong thing." In 1973 the couple remarried. "She understood my sickness more, and this medicine calms me down a lot. It helps me." In February 1974, when I talked to him, Sayre was taking heavy medication—103 milligrams of tranquilizers daily. He said he still suffered symptoms from the carbon disulfide.

"Some days are worse than others. It seems like my mind is foggy or something. I want to argue with somebody, I can't get along with my wife. I don't want to do it, but—if it's bad—it seems like there's no control there."

It seemed, when I talked to Sayre, that he was having a good day. He told his story quietly, and with assurance. He said he had asked his psychiatrist what had made him sick—whether it was the carbon disulfide. "He said he hadn't determined what had caused it." Sayre believed the doctor was reluctant to blame CS_2 because the doctor and a psychiatrist retained by the company both work at the same hospital. The company's spokesman, Dr. Charles C. Weise, said at Sayre's workmen's compensation hearing in 1971 that he believed Sayre's symptoms did not come from the chemical exposure but "reflects a lifelong state and is not due to anything recent."

Yet the company, while contesting Sayre's claim for permanent disability, readily agreed it was liable for temporary disability during the time in 1969 when he was off work. Compensation was denied.

Sayre was a young man, in his early thirties, and in spite of his problems, he viewed the future with optimism. "I think it's going to be all right," he told me. "I've learned a lot over this year. I have more respect for other people. I don't want to see them go through the same thing I've been through."

John Niewierowski began having similar problems when he worked as a cutterman in early spring of 1972. "I was dizzy, I had numbness in my legs, I'd get unbearable headaches—there's an extremely high noise level there too. My wife, she couldn't understand, she'd have a meal cooked, I'd come home, I wouldn't feel like eating, I was real touchy. I attributed it to fumes. I knew it was work-related because I had never had any trouble before. I'd seen these things happening all around me, guys going to

Highland Hospital out of their head. I'd lay in bed at night. I'd be scared, afraid to go to sleep, afraid I was going to die. I'd get pressure in my arms, my heart would beat real fast, then it would stop. My vision was affected. I doubted people, didn't trust people. I was frightened. I was in so much pain, so many things were happening. It was scary. It was frightening."

At first, Niewierowski went to a private physician, who gave him tranquilizers and told him his symptoms were not related to carbon disulfide. "I just said, 'The hell with you.' After I kind of got myself together the best I could, I went to see Dr. Buff." Dr. I. E. Buff, a crusading physician, was renowned throughout the coal fields for his efforts on behalf of miners disabled by "black lung." "I was trying to tell him what I was working with," said Niewierowski. "He said, well, it sounded like CS_2 poisoning."

Niewierowski and another worker with similar problems wrote a letter to the National Institute of Occupational Safety and Health, asking for a hazard evaluation of the plant. NIOSH investigated in May 1972. They found excessive levels of carbon disulfide and symptoms, especially among the cuttermen, of carbon disulfide poisoning. Surveys by a federal team in the spring and fall of 1972 found that twenty of the twenty-eight cuttermen interviewed and thirteen of twenty-seven spinnermen complained of symptoms consistent with carbon disulfide poisoning. Chargehands and viscose workers were also exposed, but less severely affected. "On the basis of substantial medical findings . . ." the researchers reported in November 1972, "and preliminary environmental concentrations of carbon disulfide measured, it is concluded that a *definite* and *serious* hazard to the health of workers, especially cuttermen, exists in operations conducted in the Staple Department." The health officials recommended use of respirators for exposed workers, "immediate attention . . . to both the inadequate ventilation existing in the cutting machines and general area," and the development of monitoring and medical programs.

A NIOSH report issued a year later made it clear that the

hazards continued. The report concluded that "exposure to carbon disulfide vapors at the concentrations found in this work environment is toxic to cuttermen, chargehands, or others working in the cutter area. The overall respirator program in effect for cuttermen was believed to be ineffective in protecting this group."

Niewierowski's concern about CS_2 hazards in the plant was justified, but he was not satisfied with the studies. He said he would like to see research into health effects on workers exposed to the fumes. "I'm still disgusted. I want action," he said. "This thing must be so deep it must reach into other plants. I've asked the international to provide us with a doctor. I suggested it to Stetin. [Sol Stetin was international president of the Textile Workers.] It goes deeper than just the man. It touches in his family. It hurts a lot of people in a lot of ways. It's just been hidden a few at a time. These things have been happening over a period of years and this has been swept under the rug.

"I get times I'm real rotten. I get times I can't rest, I can't sleep. I know I'm not over it. My wife understands now, she reads with me—reads the books. The NIOSH investigation helped. But I'd rather be working in a Burger Boy than working in the plant with the symptoms I've got now. Poor people're coming to work, but they don't know what's happening. They're just trading their health away for profit."

Jim Weeks taped an interview with two union stewards, Niewierowski and one named Oren Christy. The men explained how the company dealt with the problem:

Niewierowski: "If a new man shows any symptoms whatsoever, and you're still in the probation period, you get discharged."

Christy: "During that probation period, during that six months, if you show any kind of problem, if you can't hold up under that pressure, then they get rid of you. And one good strong whiff of that stuff—which you can get in one hour or one night—it can get a man right and he's had it. I've seen that happen. I've seen a man up there in the spinning area get a good whiff of that stuff and it knocked him out, colder'n a wedge . . . And they're always hiring new men up there. There was two new men up there yesterday . . . About once a week, on the average, there's three or four new men on the job, they're hiring almost one a day to fill the jobs."

Niewierowski: "I've seen as high as thirteen or fourteen come in there at once."

American Viscose officials in Nitro refused to talk to me. They referred me to Dr. Jan Lieben, medical director for the corporation in Philadelphia. "We've had some problems there," Lieben said. "There've been several cases of acute CS_2 poisoning which are all back at work and in good shape." He said the company became aware there was a problem only after the first worker was hospitalized in 1969. "Several men got ill over a period of several months. After that we went in and checked this carefully. We found these people probably had CS_2 poisoning."

I asked him when the company began to take action, and he said a respirator program was planned and a urinalysis program under way shortly before the NIOSH investigation began in 1972.

Why hadn't that been done in 1969?

"It takes a while to get things moving. And nobody has really gotten ill from 1969—a couple of questionable cases since then— the condition was not that serious."

Ventilation changes, he added, take more time to design, though he admitted "things haven't been done quite as fast as they should have been." He termed the NIOSH report "a lot of nonsense."

I asked him about workers' charges that the company has fired men who show symptoms of CS_2 poisoning. He did not deny it.

"There are people more sensitive to CS_2 than others. There are others who refuse to wear respirators."

"And these men are fired?" I asked, incredulous.

"If they are hypersensitive to that, yes."

"Well, how can you tell the difference between hypersensitivity and a simple case of dose response?"

"Only a small percentage are affected. The rest—ninety-five percent—are still in there."

"But you've just told me it's an area of high turnover."

"Certainly," he replied, unabashed. "It's an unpleasant job."

OIL

Dealing with Men as Cattle

The modern oil refinery is the biggest chemical plant of all—in fact, thousands of chemicals are petroleum derivatives—and it carries the potential for tremendous disaster. It is a potential that modern refinery practices have increasingly ignored, according to workers, engineers, and even industry trade journals. An editorial in the *Oil and Gas Journal,* April 1968, described the problems well:

> Refinery developments over the last year or so raise the nagging question: Just how far can management go with economy before quality suffers and planned savings go down the drain? The oil industry has made household gods of economy, cost cutting, and efficiency in recent years . . . The early savings in capital and operating costs were substantial. What's happened in processing over the last few months, however, is startling. New plants and units have been plagued with an assortment of start-up troubles,

106

both minor and major. Some encountered vessel failure or equipment breakdown. Many strained to get new units up to design operating levels. A few wouldn't even run. The industry also has been embarrassed with an unusual series of upsets even after plants finally went on steam. Heavy repair costs and—more important—loss of production for long periods resulting from the troubles have run into the millions . . .

Plant designs have been stripped too severely. The stress on economy has reduced flexibility of operation to the vanishing point. This situation has the earmark of a vicious circle in which quality of plant and equipment ultimately suffers. It's time management restores a measure of quality. To do otherwise will lead only to loss of the industry's vaunted efficiency and to unsafe plant operations.

A Houston engineer who helped design refineries for Brown and Root, a national contractor, was fond of stating the problem more simply. He would point them out to visitors, and say, "See that refinery? I designed that plant and it's going to blow up!"

"Sometimes they cut corners for economy's sake," he explained to me in a phone interview, "which means they cut into the safety factor. So instead of a factor of fifteen or twenty percent you have only ten percent, and this is why these things do explode. And when they explode, workers are maimed and killed."

The engineer, who asked to remain anonymous, said that during construction, short cuts take place all along the line. "The engineers usually were very conscientious," he said, "but we always felt we were playing a secondary role to management. They say, 'Get the goddamn thing out and don't worry about it. We've got so much time to make so much money, and we'll just throw it to the men on the floor, no matter what it takes.' This results in short cuts of all kinds, in material, design criteria, checking of drawings and specifications.

"The drawings go to the customer's representative—the customer always has a representative on the job. But he also is under pressure from his superiors."

Another major problem, he said, was a lack of qualified personnel. "A lot of people working for me had no background in structural design. There were many things that we let go that I just couldn't help, designs I couldn't adequately check. Time is a great factor. You just do what you can and hope for the best."

When I interviewed him in 1973, the engineer said his experiences with refinery design ended several years ago, but, he added, "It hasn't changed. It can't change. It's in the nature of the beast."

Workers tell of further short cuts. Lou Mariani, an official of the Oil, Chemical and Atomic Workers Union and a worker at Texaco Oil in New Jersey, told me in detail of cost-cutting procedures that further reduce the margin of safety. "Units that normally would be shut down for test and inspection every year to do a complete overhaul, they now try to get a three-year run on it," he said. "So the unit breaks down and emergency conditions exist—consequently people are always working on sections while the rest of the unit is working." Because maintenance workers must work on "live" systems, they are exposed to dangerous gases such as hydrogen sulfide and hydrocarbons. "Running time is the main thing," said Mariani. "Men are getting exposed more and more to unnecessary gases. We're running at full capacity all the time. Production is the main thing in a plant manager's mind. If he can keep that thing running, he's not going to shut it down. The old established practice in a refinery, they would shut that thing down. They never used to weld on a unit while it's running. Now they do it as a matter of business. One fellow was burned real bad because of this. He was in the hospital over a year.

"Now the argument that management will say: 'Well, you're working in a refinery.' But I've worked in a refinery for twenty-two years and the fellow that I'm working with has for thirty years, and we never had atmospheres that we've had for the last fifteen or fourteen years.

"The pump rooms, they run 'em until they collapse. You can't walk in these pump rooms any more because of the gas. You just can't stand it. A fellow working in there, because of the gaseous atmosphere, he was overcome. They thought he was asleep.

"On the catalyst side, where the catalyst goes over to the gas line—it's a forty-eight-inch line—our men have to go in and patch it. The men are working under this atmosphere. When they had to shut this unit down a year and a half ago, because of the conditions I told you—the cyclones went bad, when they opened up that unit you could actually see through the cyclones. They're supposed to be solid steel. They were completely eaten away.

"The crude oil units used to shut down twice a year. Now they run it for two years. They had a fire a month ago. A pump had been written up and work orders put out for it because of vibration. So what happened, they had trouble with the pump, had to shut it down, and put a spare pump on—a spare not as large and not a hundred percent efficient because it hadn't been overhauled. They decided to try the other pump again. It had a leak. First it leaked out all this gas into the atmosphere, then sparks, and Boom! a fire. It burned out the whole pump room and part of the controls. That fire could have been prevented."

Every major refinery in the Delaware Valley has had a fire of major proportions, Mariani claimed. Certainly fires and explosions have become commonplace in refineries, and with such engineering and construction practices and work practices, it is no wonder.

On October 11, 1971, the OCAW local at Mobil Oil in Paulsboro, New Jersey, requested an imminent danger inspection under the Occupational Health and Safety Act of 1970. The result was a highly controversial inspection by OSHA lasting twenty-eight days, and resulting in citations for 354 violations of the act. Union officials, however, were not satisfied with the inspection. I decided to visit the plant to find out why.

The plant, originally built by the Vacuum Oil Company in 1916 on one thousand acres of farmland along the Delaware River to produce five thousand barrels a day, gradually expanded until in 1973 the refinery handled a million barrels of crude oil a day and employed thirteen hundred employees. It is one of the largest refineries of "one of the ten largest industrial corporations in the world," by Mobil's own account, and produces "more than five thousand petroleum and chemical products."

The Paulsboro plant looked as I expected it to look. It was a huge complex of pipes and pumps and vessels and smokestacks, a macabre industrial sculpture of steel set on a flat swampland. I met with company officials in the office of plant manager Ron Niederstadt. Joe Mouton, employee relations manager, and Robert Wiener, a public relations man from Mobil's corporate office in New York, were there to assist Niederstadt, who looked the very picture of a successful and rising business executive—handsome, poised, confident.

I brought up the OSHA inspection, and Niederstadt, with the executive's ability to create his own logical universe, pooh-poohed the 354 violations and expressed his satisfaction with the way things had gone. "They gave us a fairly thorough inspection," he said. "After these two tests I feel fairly comfortable that I have a work environment that's fairly satisfactory—that I've been checked and not found wanting."

But how about the citations for rotten stairs and walkways, which OSHA inspectors termed a serious violation? I asked.

"We're continually repairing," he responded. "If an unsafe condition develops, then we correct them just as quickly as we can."

But, I pressed, how can you say that, when conditions have been allowed to deteriorate to the point that materials are actually *rotting?*

"If an unsafe condition develops, then we correct it just as quickly as we can."

Questions and answers followed that general line for several minutes, until Niederstadt, exasperated, suddenly jumped up and ran to the full-length glass windows looking out on the refinery. "I work here too!" he exclaimed, gesturing toward the windows. "If it blows up, I go too!"

A few other samples of Niederstadt logic:

On the subject of overtime: "We have voluntary overtime. Occasionally we've forced people when we ran out of volunteers."

On a fatal accident: "It was due to human failure. OSHA indicates the failure was on management side. We didn't contest it. It was the fault of the union-represented people that did not follow instructions."

Union officials explained the deficiencies of the inspection to me in patient detail. They believed that OSHA inspectors missed the most serious problems in the plant because each day they presented management with a schedule of which operations would be inspected. That gave management a chance to cut back operations temporarily, the union officials said, to clean up and to generally hide evidence of problems. Another problem noticeable in most OSHA inspections, they pointed out, was that the inspectors tended to concentrate on safety and avoid health problems.

Niederstadt expressed relief that none of the violations cited were for serious health problems. The company provided me with this statement: "The OCAW official claimed there were nine 'imminent danger' conditions at the refinery, and that many more safety and health hazards existed . . . The nine so-called 'imminent danger' situations *alleged* by the union were general in nature and had to do with health rather than safety. These allegations covered such items as asbestos pipe coverings; sulphuric acid mists; overexposure to lead, caustic soda, phenol aromatics and ketones; mercury problems with instruments; reports of deafness in the filter plant; excessive carbon monoxide in compressor areas, and severe noise problems . . . In fact, OCAW was not upheld on any of its allegations that nine "imminent danger" conditions existed at the refinery."

Mobil, one of the richest corporations, was fined a total of $7,350 for the 354 violations, or an average of twenty dollars per violation.

Union officials continued to maintain that the conditions had existed and continued to exist. And after talking to a number of workers, I had to believe the union.

Steve Wodka, legislative representative in the OCAW's Washington office, said the union filed the request for the inspection because Mobil had refused to recognize the union safety committee. This was the first time the union had requested the inspection of an entire plant. The union was dissatisfied with the inspection before it even began. "We filed for an imminent danger inspection and it took two weeks for OSHA to respond," said Wodka. "Then, we just saw one thing after another. The inspectors were overly cautious, condescending to workers and overly submissive to the company. The inspectors felt that 'you guys really don't know what's good for you.'"

Wodka said the union was not satisfied with the results, either—the small fine for 354 separate violations. "We felt it was inadequate," he said. "Only eleven were on industrial hygiene, which we thought was the biggest problem."

In May 1972, when a fire and explosion resulted in the death of a worker, Mobil was fined $1,215. The union felt it should have been more. "The fine for a willful violation is ten thousand dollars," said Wodka. "OSHA refused to declare it is a willful violation. We have no legal recourse. You cannot protest the weakness of a citation or the amount of fines. The only thing you can contest is abatement, while the company can contest the whole thing.

"In October of 1972 there was another fire and explosion. A man trapped under the rubble managed to crawl out safely. So we had to pull OSHA's leg again. The regional administrator wanted to drop the whole thing. 'When are you guys going to wrap it up?'— that's what he said. We said, look, you don't understand. The nature of the problem is that it's a continuing problem. Well, they didn't get out there until [a few days later]. All the rubble has been cleaned up, the blood's been washed out and the dust has settled. So we were really surprised at the citation—a nine-hundred-dollar fine.

"The law was written that the Secretary of Labor would be a benevolent dictator on behalf of the workers, and that hasn't turned out to be the case."

The May accident, resulting in the death of George Kugler, a tube-still operator, could have been prevented, union representatives told me. Harry Bailey, a union steward, said he had mentioned the hazardous situation to OSHA inspectors earlier, "but there was never anything wrote up about it."

"Kugler was down checking the pump at the dehexanization," said Bailey. "Every time they would do that they'd—it's like everything in there, they never repair anything right. The other fellow was up the other end. She flashed and he was burnt to a crisp. That's all we know. When they found him, they picked him up with a shovel. The safety man wanted them to take him to the hospital. The company has this idea nobody dies in the plant. They want them to die on the way to the hospital.

"The company dismantled the pump and removed the nipple. The flash burnt the substation wires across the road. They scrubbed that before the federal men ever came in there."

The inspection, said Bailey, "wasn't what we wanted. We have more severe hazards where people were dying down in this plant, and we wanted to know why they were dying. Anytime you have anything wrong on the inside of you, they classify it as a personal illness. Even senators are working against us down there. They're only interested in what those federal inspectors write down. Why don't they write down that the company forced men to work 'doubles'—two shifts down in ammonia? Tears were coming out of their eyes, they couldn't breathe and the company was forcing the men to work. Management says anything that interferes with their profits is an emergency.

"They used to bring a unit down once every six months, then it went to a year, now they're trying to run it three years. When the unit isn't running right, they say, continue running it on a calculated risk. If the unit burns up, well, it's an unavoidable accident and the insurance company rebuilds it for them. The biggest problem is a lack of repair work around there. They don't keep maintenance up."

One safety hazard the union had complained about was an area where a "mule" hauling four cars, each carrying fifty-five-gallon drums, came down a ramp, then had to turn and go up another ramp. "You have no control of that mule coming down the ramp," said Bailey. "You just have to ride it out." Bailey said he showed the area to a federal inspector, but the company man said when conditions were bad they cut down the number of cars. The inspector did not cite it as a hazard. A few weeks later, Joe Mondile, a Mobil worker, was walking up the ramp as a mule coming down the ramp slipped on oil, jackknifed and hit him.

"My back was turned," said Mondile later. "It didn't make any sound and it pinned me against the wall. It twirled me around. Then I seen the drums against my chest. The wagon is what got my leg." The muscles of his chest were ripped away from the rib cage and a long gouge was torn in one leg.

When he was hit, Mondile landed in a puddle of water. There had been a leak in the pipe at that spot for two years. "Two years employees have been writing slips on that leak," said Mondile. "Within half an hour after they put me in the ambulance and took me away, they had a pipefitter over there to fix that leak.

"I find it's a company that preaches safety but doesn't practice it. Their biggest excuse is they're always in the process of fixing things. If it's production it gets fixed. If you want something done for safety, it's 'No manpower.' They could use another thousand men for a year—that's just for cleaning up and repairs. There's too much overtime, and a lack of maintenance and housekeeping and a 'I don't give a damn' attitude as far as management is concerned. How many barrels of oil have been refined today and that's all they're interested in."

Mondile was forty-nine and had worked at Mobil for twenty-five years. He took the job because "it had security," he said. "That's a big thing when you're a family man. Good benefits. A married man with children thinks about these things."

Union president Dick Meyer said the union had difficulty getting men in the plant to cooperate during the OSHA inspection. "Some people know their unit like they know their wife, I guess, but a lot of places we couldn't get anybody that would talk. The history of the oil industry is your grandfather worked there, your father worked there. For years it was a good plant. We used to hate to miss a day. All that stuff changed in the fifties." The average age of Paulsboro plant workers was fifty-five, he said, and most of the men continued to believe that the company was looking after them. "I think if they'd cut 'em a dollar an hour, they'd say thank you," he said.

"They win safety awards pretty near all the time," said Meyer. Like many plants, he said Mobil kept its lost-time accident rate down by bringing injured workers back into the plant. "One time my leg swelled up from a jackhammer. They sent me to a doctor. He put it in a cast. They brought me in there. There was a man named Green that got burned pretty bad. I saw him one night—he was crying. They brought him in there in an ambulance so they wouldn't get that lost time. That's been going on for twenty years. I can't tell you how many people's been coming in there in pain."

It's easy for the workers to talk about accidents that are readily recognized as hazards, but they generally are not nearly as familiar with the chemical exposures. "In the ketone area where ammonia is," Meyer went on, "people started falling out—and you wonder what's happening. When people were starting to die in their forties and fifties, that's when I started to get concerned."

One chemical hazard common to oil refineries is hydrogen sulfide gas. The gas is a by-product of the refining of oil, produced by the combination of sulfur in the crude oil with hydrogen. Pipe fitters at Mobil consider it a major hazard, with good reason.

Don Fuller, a Mobil pipe fitter, had worked for the company thirty-two years at the time he was gassed.

"The day it happened, I'll never forget," he said.

"I was kidding Eddie, my boss. He had his arm wrapped around me. He said, 'Guess what, Don?'

"I wrapped my arm around him. I said, 'Guess what, Eddie? My wife told me how much she loved me this morning. I want to see her tonight.' I said, 'Don't hurt me. It's my wedding anniversary.'

"He said, 'Look up there—that's where you're going to be working.'

"I said, 'Oh, no, Eddie, you don't mean we're going to be working on that gas line—that hydrogen sulfide?'

"He said, 'It's all right, it's been cleared.'

"I said, 'All right, Ed.'

"There were about five men in our gang. For some unlucky reason, I climbed the stairs first. The next guy that came around the railing saw me and screamed. They thought I was dead."

Fuller was taken unconscious to the hospital. He remained unconscious. "One doctor, Snider, my wife said, he stayed there the longest. He'd say, 'If only he'd come to,' and he'd look and feel and then he'd go out. He said, 'If only he'll come to screaming and fighting, everything's going to be all right.' He said, 'If he comes to like a baby and just lays there,' he said, 'he won't be no good.'

"So she said it was the twenty-seventh hour. She said, 'You opened your eyes and started screaming and fighting. She said, 'They had to tie you down.' "

At first Fuller didn't know who he was or where he was, and even after two weeks he thought it was 1955, not 1972. Eventually, with his doctor's help, he regained most of his memory, though he began to forget people's names. "When I see a man on the street I've known thirty-two years I go through everything trying to remember his name. I get embarrassed, so I duck in a store. Damage has been done somewhere."

Headaches were Fuller's biggest complaint. "One night lying in the hospital, I'd swear somebody stabbed me right in the head.

Ever since then it's headache after headache. I keep a record every day—date, headache, date, headache, date, severe headache, date pretty good, date, headache, all the way through. I say to my wife, 'Get my belt and tie it around my head tight as you can.' 'I will not, you'll kill yourself.' I say, 'Betty, I get relief.' But I can't get any relief from it."

Joe Duca was another Mobil pipe fitter. He worked for the refinery forty years and told me he had been gassed with hydrogen sulfide several times.

"Each time I got hurt and fell off the scaffold. The last time was three and a half years ago. Seven years prior to that me and seven other fellows were gassed. I fell on a steam line six feet below. Five of the other guys did too. One guy got his neck all messed up—Joe English—because he fell on the cement.

"This last time we were taking a pipe line apart. We knew the hydrogen sulfide was there. The safety man came around and checked and said it was safe to work. But they didn't order us no gas masks.

"When we took the flange apart and put the blank in, it was okay. But maybe somebody neglected to check the other end. When we finished the job, I took the blank out. I was alone on top of the scaffold. I don't remember what happened. I just passed out.

"For three weeks I was just like a dead person. I couldn't make my voice come out like I wanted, say things I wanted to say. Even now sometimes I can't remember things. I forget awful quick. There's more to that than I can say. I'm still going to the doctor and taking pills for my nerves and my head. And I have a lot of noise in my ears—like a generator—all the time. My head sometimes doesn't feel good. I get headaches. Another thing, sometimes I get dizzy, too. I can't balance myself. I get in a sweat, like, and I know it's coming and have to sit down.

"After that last gassing they wanted me to work. I worked about a week and a half. My nerves got so shot they took me back to the hospital for two weeks. After that they told me I better take my pension.

"I hope somebody puts something in there that wises the company up. So they show some concern for the men. It used to be like a family, now it's dog eat dog."

Les Jandoli, a Camden, New Jersey, attorney, handled about thirty compensation cases a year from the Mobil plant. "The problem down there is they have a very weak union," he said. "Mobil discourages men from filing claims. I've seen letters from supervisors: 'If you continue to lose work you may be let go.' After thirty-two years. A fellow came in last week, he'd been off nineteen days over and above what they allowed for sick leave and they threatened to fire him. It's happened time and time again.

"A lot of men develop lung problems down there. All the pipes are wrapped in asbestos. Pipe fitters, welders, et cetera, are submitting themselves to a hazardous condition. The men are not willing to strike for better working conditions, not willing to put their jobs on the line for each other. Gloucester County was a farm area—now, they're making a nice standard of living which their fathers and grandfathers never did."

The company thinking, Jandoli explained, was this: "They're paying the men and that's all their responsibility extends to. They feel as long as they do pay the men, their responsibility is over. They are so used to dealing with men as cattle, when the men do raise issues, they can't cope with it."

Jandoli added, "Mobil isn't by any stretch of the imagination a bad employer."

AUTOMOBILES I

The Murder of Foreman Jones

> Your technology hasn't advanced to the point where you can do without the human element. No, the human element is something that has to be considered too. Like a machine needs oil, and the motor needs gas, that human machine needs a little bit of kindness and compassion and understanding too to function properly.
>
> —*Eldon Avenue worker*

Sometimes at Chrysler's Eldon Avenue Gear and Axle Plant in Detroit an overhead conveyor runs away. The overhead conveyer is like a mechanical clothesline on which iron parts are hung and carried from one department to another, above the tops of other machines and over the heads of the workers who tend them. Like most machines it requires regular maintenance. Maintenance is a

119

crucial function in any factory, necessary not only to keeping the machinery operating, but also for safety. Maintenance is expensive, and between the effort needed to sustain production and that to assure safety is a wide and costly gap.

If a plant is running on a tight budget—and most do—it will try to cut costs on nonproduction items. Regular overhauls and inspection of machinery will be cut back. Maintenance of safety guards, ventilation systems, lighting, railings, ladders, restrooms, fire-fighting equipment, and so on will be placed low on the list of priorities, and such items may remain in a state of perpetual disrepair. And so it occasionally happened at Eldon Avenue that maintenance of the overhead conveyor was too long neglected. A crucial part would fail, and like a car without brakes, the line would run away with its load, rushing heedlessly on, hurling off dangling pieces of steel in all directions.

Sometimes, however, it is the man who tends the machine who has been neglected or pushed beyond the point of his endurance, and he too may rush machinelike out of control. The worker who threatens his foreman, the worker who "speeds" on amphetamines to keep up with the machine, or who takes barbiturates to dull the pain of meaninglessness, the worker who beats his wife, who fights in the bar after work, the heavy drinker—all of these may be reacting to the pressures of the job, the stress of hazardous work. These stresses can be overwhelming if the worker sees no way out, no alternative but to accept conditions which he sees will lead inexorably to his destruction.

One midsummer afternoon in 1970 a black man named James Johnson, Jr., became the symbol of the auto workers' rage and despair, a hero to the powerless, the "equalizer" who caught the shit and threw it back. But from start to end the only role he really played was that of victim.

He was the son of Edna and James Johnson, Sr., sharecroppers on a Mississippi plantation. His mother was fifteen when she married in 1933, and was in the fifth grade at Josie Creek school near

Starkville. His father was in the seventh grade at the same school. James was born a year later, the first of five children. The family lived in a two-room hut, a bedroom and a kitchen, a bed in each room, without plumbing or heat. They shared an outhouse with other families on the plantation. Five months a year the children went to school; the rest of the time they worked with their parents on the plantation, and though everyone worked hard, there was never enough money for food or clothing. The children went without shoes and coats. Mrs. Johnson sewed rags together to make quilts to keep the family warm on cold nights. An average Starkville winter brought at least one snowfall, and the Johnson house, which was not sealed, was always cold.

As a child, James Jr. was quiet and kept to himself. Other children teased him and called him a sissy because he would not play with them. Death came frequently to the little plantation settlement of about fifty people, and the deaths left a deep impression on the boy. Even when he was four years old, when his great-grandmother died, James seemed to react with a great fear which continued throughout his childhood. When he was about ten years old, he saw his cousin Henry's mutilated body lying on the highway outside of Starkville, cut and dismembered by a mob maddened at the young man's romance with a white girl. The whole family "took it hard," a cousin recalled later, but it especially affected James. Shortly after that he began to have "spells." He would hear voices and see distorted faces of the dead—people he had known—beckoning him to join them. He would sweat and moan and cry out that he was going to die. His mother would hold him in her arms and try to comfort him. She did not know what a psychiatrist was, but she worried about him and took him to the doctor in Starkville. The doctor gave him medication for his "nerves," and said he would probably grow out of it. Mrs. Johnson thought the medicine seemed to help, but the spells continued.

During this time James's father was in the service, sending money home every month. With eight hundred dollars she was able to save, Mrs. Johnson bought another house, hardly better than the first one except that it had three rooms and a private outhouse. Mrs. Johnson found a job as a domestic servant in Starkville, earning twelve dollars a week. Town was five miles from the Johnson place, and she had to walk or hitch rides to get into work. At first she came home every night, then only on weekends. Finally she just gave up coming in at all, leaving James Jr. to cook and look after his younger sisters and brother. When James Johnson, Sr., returned from the Army, the family continued to have troubles with money, and the Johnsons' marriage began to fail.

In 1953, when James Jr. was nineteen, his parents sent him to Mt. Clemens, a suburb of Detroit, to live with relatives and finish school. He went to Mt. Clemens High School and had a girl friend for a time—they would be seen walking to class together, holding hands—but after a while, Johnson began keeping to himself, avoiding girls and parties completely. Since he came north his "spells" had stopped, but he began to have headaches and would go to his room, saying to himself, "James, you're so stupid and ugly; no one wants to see you; go hide under your bed." * His aunt would find him lying nude on his bed, or sometimes under it, swearing to himself, sweating, and seemingly in a daze.

For all the Johnson family, Detroit stood beckoning, promising a way out of the poverty of southern plantation life. You could get an auto job, work your way up in the plant, make enough money to have a nice house, a car, and raise a family. These were the hopes and dreams of the Johnsons and thousands like them. They came from the Appalachian coalfields, the Carolina mill towns, and the sharecropping fields of Georgia and Alabama and Mississippi, percolating north through the factories of Dayton and Cincinnati toward Detroit, looking for the coveted jobs in the auto plants.

* Much of the material about James Johnson comes from notes his lawyers made in interviews with Johnson and his family, as well as from court testimony and the workmen's compensation decision.

For many, the dream proved to be an elusive one. James Johnson, Sr., had tried his luck in Detroit in 1956, after he divorced his wife. He worked construction, making less than a hundred dollars a month, always hoping for a job at "Ford's" or Chrysler, but he didn't get it. He stayed for two years, then returned to Starkville to the job he had left behind—earning fifty dollars a week as a janitor. James Jr.'s younger brother also had come to Detroit and also failed to get an auto job and returned south. Only James Johnson, Jr., made it. After serving in the Army, he returned to Detroit, held several low-paying jobs; then finally, in May 1968, he hired in at Chrysler's Eldon Avenue Gear and Axle Plant. He was thirty-four years old, and earning eighty-six dollars a week after taxes, seventy-eight dollars after union dues. It was not a lot of money, but it was the most he'd ever made.

Johnson was a conveyor loader on the No. 2 oven in Department 78. There were two conveyor lines moving through the No. 2 oven. Johnson stood at the foot of one of the conveyor belts and Joe Andrews at the other, and by hand they unloaded the brake shoes as they came, ten seconds apart, hot from the oven. The men wore specially lined gloves, yet even those would wear out in eight hours from the rough metal, the heat, and the grease. The plant was not air-conditioned or well ventilated and on a hot day the temperature in the oven room might reach 120 degrees. Many workers considered it the worst job in the plant.

Andrews had a month more seniority than Johnson. Before long he moved up to a better job. But Johnson stayed at the ovens eight months before he was transferred to a better job in the cement room. Even though he had a new job, his classification and pay remained unchanged. He was disappointed, but he continued to work hard. He seldom spoke to other workers, did not loaf, and was always on time when the shift started, always the last to punch out at the end of the shift. This was the best job he'd ever had and he wanted to hold on to it. He bought a secondhand car and a house, which he shared with his sister Marva, sent money home for his mother and younger sisters, sang in the choir at the Calvary Baptist Church on Sunday, and dreamed of the future.

During the spring of 1970, Eldon Avenue was gaining notoriety throughout Detroit. Conditions in the plant had been deteriorating for several years. Eldon manufactured the complete gear and axle assembly for Chrysler cars. It was a noisy, dirty process made dirtier and more hazardous by inadequate maintenance. Plant manager Harry Englebrecht admitted to some safety problems in a taped interview: "The plant had reached a period—and this goes back way before the Johnson incident—where the requirements of the plant exceeded the capacity. We had to produce the axles for the entire corporation. And there was an interim period when the corporation demanded more axles than this plant could efficiently produce, and as a result of that, the conditions became crowded, with more equipment than normally you would put into this plant . . . So, during this interim period, when we were asked to produce more than we normally wanted to produce, the plant became overcrowded. It becomes difficult then to maintain some of the standards you would like to maintain in a plant . . . There are limits as to what maintenance you can perform when you're running on an all-out basis."

Englebrecht did not specify which standards became "difficult" to maintain, but Eldon Avenue workers listed the following in interviews: Fork-lift trucks or "jitneys" were in a constant state of disrepair, light fixtures often fell down, aisles were crowded, skid boxes full of parts were stacked dangerously high, oil was everywhere. "Oil on the stairs to the johns," said a former union steward, "oil seeping up from the floors, oil coming around on the floors from the machines, oil on the racks you have to stand on while you're operating the machines."

Ventilation was another major problem. The air was filled with a mist so thick that it was rare to see from wall to wall. The mist is caused by evaporation of a coolant—a mixture of oil and water—

used on the grinding machines. This milky blue liquid pours continually over parts of the grinder, cooling it as it cuts and grinds.
The liquid collects in a large tray surrounding the machine and
sometimes spills over onto the floor as well. Few plants engaged in
heavy industry are a pleasure to work in, but the gear and axle
plant, with its constant screaming noise, black greasy towering
machines and conveyors, black greasy supporting beams and walls,
dim lighting, oily floors, and the heavy blue mist which hung over
all, was particularly grim.

Accidents were common. On an average day, according to two
former company attorneys, ten to twelve workers were injured
seriously enough to be sent to workmen's compensation lawyers at
Chrysler for evaluation. In a plant population of three thousand
workers, more than one serious injury per person every year—a
lot more than ever appear in official statistics. Statistics list only
"lost-time" accidents, and Chrysler, like many companies, strongly
discouraged workers from taking time off. "Like one man got his
finger cut off," says Jordan Sims, a former steward at Eldon. "And
they said, 'Well, it's not like losing your whole hand. You're not
totally disabled. You got one hand left. Come on in.' And once he
came in, see, they wouldn't tell your line foreman about your PQX
[specification for light duty], and if you didn't have no special
PQX, and sometimes if you had one, the foreman would say, 'Well,
this is the only way I can use you. You take this, or nothing.' And
nothing usually means you're going to be disciplined if you don't
take this. And we've got people out there who are hobbling around.
They have to hobble to the job on crutches and maybe sit there and
lean against the line to do their work. Chrysler does this all the
time, trying to protect their lost-time record. This is standard."

Joseph Baltimore, a former workmen's compensation adjuster
for Chrysler, now an independent attorney, corroborated Sims's
description. "We had a number of cases where people have operations, fingers cut off, and they bring them back the same day. It
just doesn't make sense, but the statute allows you to do that and

the employers take advantage of it." He described the case of one woman, Mrs. Rose Logan, who was struck by a fork-lift truck, a "not uncommon" accident, he said, "because they have a quota and the driver has to keep moving. There should have been a rail for protection, but they only had a rail at certain places." Mrs. Logan, he said, was brought back into the plant in a wheelchair, so management could avoid paying her compensation. "Chrysler found her favoring work to do—dusting in the offices, folding towels—all of this while she was in a wheelchair," says Baltimore. "I didn't like it. It was my feeling that a person with a broken leg should not be in the plant at all. I told her if any day she didn't feel like coming in that we would pay her compensation."

Another lawyer who worked as a compensation representative for Chrysler, and asked that his name be withheld, said: "There was a lot of pressure on the workmen's comp rep to turn down cases. Mostly from the plant safety man, who would always try to find some way, you know, of getting out of the case so he wouldn't have a lost-time." His experiences at Chrysler have convinced him that safety is impossible at Eldon or in any plant because "capitalism requires profits and the only way to keep on making profits is to speed up, cut down on expenditures.

"Workmen's compensation is a liability, considered a nonproduction expense," he said. "At Chrysler, a conscious decision was made—I used to participate in this decision-making process—'Do we spend more money on workmen's comp, or do we spend more money on safety?' Now when I was working in workmen's comp, at the year end, what we had to do was figure what they call the 'reserves.' Chrysler puts aside a certain amount—I think it's ninety-eight cents per hundred dollars of wages they put aside for workmen's comp expense. What we had to do every year was go through all of our cases that were open and estimate the workmen's comp expense for the coming year on those cases, and turn that figure in to accounting. And they did their thing with computers and figured out what the reserves should be. How much total money it would cost the corporation in workmen's comp, and balance that against the safety department estimates for each plant, you know, how much it would cost to make the plant safe. And it always turns out that the workmen's comp expense is less. Therefore they don't spend money on safety.

"That's why Gary Thompson died. Because the fork-lift trucks weren't repaired. Because there was production pressure.

"Do you understand what a fork-lift truck does in a plant like that? They go to the end of the line, and people put parts and stuff in skids, and the fork-lift trucks come and take them away. You have to have this, otherwise you have all these skids full of parts, and your line stops because there's no place to put them. Do you know what chips are? Gary Thompson was unloading a load of chips when he was killed. There's a lot of grinding and boring done in that plant. The chips [metal waste] have to be pulled—that is, emptied into large hoppers, where a fork-lift truck comes and takes them away. Otherwise the machine clogs up. So you see, a fork-lift is essential in the production scheme. If you don't have a certain number of fork-lift trucks and they're broken down, you fix them or you keep running them when they're unsafe."

Gary Thompson, a twenty-two-year-old Vietnam veteran, was found dead, crushed under a loaded fork-lift truck on May 26, 1970, only two months after he had started working at Eldon Avenue. Lloyd Utter, UAW safety director, investigated the death. According to his report, it was Thompson's first experience driving a fork-lift. The accident occurred on his second trip. Apparently the fork-lift had rolled backward, Utter reported, crushing Thompson, who was found under the loader and the hopper with "only his feet extending." Utter inspected the fork-lift, found the emergency brake broken and disconnected. The shift lever was "loose and sloppy," the steering wheel loose, and "the equipment in general . . . sadly in need of maintenance."

Utter inspected the rest of the plant. He found the repair shop full of fork-lifts in disrepair—missing lights, brakes, horns, with loose steering wheels and leaking hydraulic equipment. He was appalled by "horrible conditions [at] the scrap lot." He recommended that action be taken to prevent "the dangerous draining of oil off the scrap onto the aisles and traffic-ways." He noted finally that "there seemed to be little attempt to maintain proper housekeeping except on the main front aisle. Water and grease were observed all along the way as we proceeded. Every good safety program has as its base good housekeeping procedures. Steps have to be taken immediately to improve conditions within this plant."

Utter's discovery was no news to Eldon workers. Thompson's had been the second death in the plant within two weeks. On May 13 a fifty-one-year-old black woman named Mamie Williams had died. Her story could not be confirmed, but it followed an often-told pattern at Eldon and other auto shops. According to a leaflet distributed by ELRUM—the Eldon Avenue Revolutionary Union Movement, a part of the Detroit-based League of Revolutionary Black Workers—Mamie Williams had had a long history of illness and had been ordered by her doctor to stay at home and recuperate. Chrysler, however, allegedly ordered her to return to work immediately or be fired. Fearing for her job, the leaflet said, she returned to work but shortly became "deathly sick . . . Management had to carry her out of the plant in an ambulance." The next day she was hospitalized, and on May 13 she died.

On May 27, the day after Gary Thompson's death, Eldon workers walked out. It was the third wildcat strike at Eldon in less than two months. Second-shift union stewards led the first walk-out after an argument between a worker and his foreman, in which the foreman threatened the worker by brandishing a pinion gear and saying he was going to bash his head in. The employee was fired, but no reprisals were taken against the foreman. The second wildcat occurred a few days later, after Chrysler summarily fired the second-shift stewards who had led the first walkout.

At this point, had Chrysler been faced with strong, solid opposition from the United Auto Workers local 961, backed by international union leadership at UAW's Solidarity House, Chrysler would likely have been forced to capitulate—to reinstate the fired union stewards—thirteen in all, the entire union leadership for second-shift workers. But local 961 officials, though they had authorized the second wildcat, later denied they had done so. Chrysler won an injunction against the striking workers and union officials ordered workers back into the plant. The expelled union stewards, denied the support of their union, remained fired. One of these men, Hugh Jones, was James Johnson's steward.

Had the union been doing its job all along, of course, the wildcats might have been unnecessary. The fact that the wildcats occurred, that at least one shift and all thirteen of its stewards supported such an action, means that they felt powerless to gain their demands for a safer plant through regular union procedures. Those procedures, cumbersome at best, were even more so because local 961 leadership, concerned primarily with maintaining power over the local, preferred to deny that safety problems existed.

"We don't really have any maintenance problems," Dan Tumer, local vice president of 961 told me in an interview a year later. "When we took office two years ago—that was in July of '70, we told them this is one of the things they had to do—clean up the shop. Hey, in any shop you're going to have something go down at any given time, and that's going to create a problem, but we really don't have any problems. The main thing is the management-union relationship. If we have a good relationship we accomplish something. If we both take a hard-nose stand then we're fighting each other. The only one that's going to get hurt is the worker." He explained workers' complaints about conditions in the plant as merely unfounded, childish rebellion. "It's human nature to rebel against your boss. It's an everyday thing. You find children at home going to rebel against their father. You take people working in a bank, you'll find their state of mind isn't too good either."

Because of such attitudes in the local UAW leadership, dissident groups flourished at Eldon in the spring of 1970. In addition to

ELRUM, a similar movement for white workers—"Eldon Wild-cat"—was active in the plant. And after the wildcats and firings, several of the disfranchised stewards formed the "Eldon Avenue Safety Committee." The committee issued safety bulletins to alert workers to specific hazards in the plant. James Johnson read the leaflets and agreed with them. He had suffered his share of in-juries—he had lost one finger and lacerated another in a conveyor belt. He also had suffered from back pains, chest pains, abdominal pains, nausea, and headache, and he blamed it on strain from work.

The committee's leaflets dealt as well with racism and harassment of workers by management—subjects Johnson also knew some-thing about.

It was the spring of 1970. Johnson had been working at Eldon two years. He was still working in the cement room and, in spite of his exemplary work record, was still classified as a conveyor loader. Robert Bayne, a job setter in the cement room, had noticed that Johnson hadn't been promoted. Bayne taught him to set up the cement machines and told him that when he needed someone to help on that job, he'd ask for Johnson. Johnson would gain ex-perience and thus when there was a vacancy on the job, Johnson would be qualified for it. Johnson's hopes seemed about to be realized.

When Bayne was due to go on vacation, he requested that John-son be assigned to replace him. But to Johnson's disappointment, the foreman gave the job to one of his friends, a white man. Ac-cording to Johnson, the foreman, Bernard Owiesny, regularly in-sulted Johnson, the only black worker in the cement room, calling him "nigger" and "boy." As the lowest in seniority in the cement room, Johnson was frequently called off his job to work in other areas of the plant, chasing stock or working on the wet line or the ovens, often without sufficient instructions on how to do the job. Owiesny allegedly delivered orders to Johnson in an abusive man-ner, telling him, "Do this right now, boy, and I mean *right now, boy*," and if after neglecting to instruct tim, the foreman found

Johnson's work unsatisfactory, he scorned him with remarks like, "You niggers can't catch on to nothing."

Early in May 1970, Johnson was hurt in an auto accident. His car was destroyed, and Johnson suffered back and neck injuries. His doctor advised him to stay off work while he recovered, so Johnson went to the plant to pick up insurance forms. Chrysler personnel directed him to the company clinic, where Johnson said he was perfunctorily examined by a Chrysler doctor who told him he was fit to return to work. Instead, Johnson followed his own doctor's orders and remained home a week, until he received a letter from Chrysler ordering him to return to work or be fired. Johnson was afraid of losing his job, so he went back.

At the end of May, Johnson was scheduled for a one-week vacation. Owiesny, his regular foreman, was on vacation, so Johnson took his time card to the substitute foreman, Hugh Jones, to get signed out, the usual procedure. Johnson thought he heard the general foreman, Elbert "Jim" Rhoades, tell Jones to "clear him." "Clear" is slang for dismissal.

The remark preyed on Johnson's mind during his vacation and he began fearing for his job. He began to have nightmares in which he was fired for going on vacation and other nightmares of parts falling on him from overhead conveyors and of slipping and falling on greasy and wet floors at Eldon. Then the nightmares began to come true.

When he returned from vacation, his time card was missing. A few days later he received a registered letter from Chrysler notifying him that he had been fired for unexcused absence. On the advice of another worker Johnson continued to work, punching in and out with a makeshift time card. He worked almost a month before he was officially reinstated. Chrysler told him it had been a mistake. Johnson did not know that the termination notice had been based on an order to "clear, AWOL," signed by Owiesny. The foreman had returned during Johnson's absence, and should have known that Johnson was on vacation—employees' vacation schedules were posted. After that incident, Johnson believed with some justification that somebody was plotting to get rid of him.

On July 16, Johnson had been working about an hour when foreman Hugh Jones came over—Owiesny was temporarily working another shift. Jones pulled Johnson off his regular job and ordered him to work the No. 4 oven, unloading brake shoes from two conveyor lines. It was the job he had done when he first hired in at Chrysler two years ago, a dirty job, and in July, with temperatures outside at 80 degrees the oven room was close to 120. There is some confusion as to exactly why Johnson refused to do the work. Johnson has maintained that he refused only because he didn't have the protective asbestos-lined gloves necessary to do the job. But he has also complained that there were others in the plant with less seniority who should do the job. He believed he was being kept at a "conveyor loader" classification unfairly, just so the foremen could order him to do such jobs. For whatever reason, Johnson refused and asked for his steward.

Charles Horton, Johnson's steward, was one of the thirteen second-shift stewards fired after the wildcat strikes earlier that spring. Horton and several other stewards had later been reinstated after they agreed to sign a statement pledging not to engage in wildcat activities again, at risk of losing their jobs. It was an effective intimidating technique, especially since some stewards, including Jordan Sims, who had twenty-two years seniority, had not been reinstated. The stewards could read the message. As a result, Eldon workers were left without effective representation on the floor. They knew it, and tensions in the plant were high.

When Johnson refused the job, foreman Hugh Jones put Johnson's former partner, Joseph Andrews, on the No. 4 oven, then called union steward Horton and Jim Rhoades, the general foreman. Johnson sat, stripped to the waist in the hot oven room, while Horton and Rhoades conferred with Jones. Then Rhoades turned to Johnson and said, "Come on, boy." As the four men walked to the production office, Horton told Johnson, "Sorry, James, I can't help you." Jones had already made up his mind to suspend John-son. In fact he had started to write up the order on the floor, even though the UAW contract called for an informal conference before a suspension could be issued.

The discussion in the production office was merely a formality. Jones did most of the talking while Johnson stood silently, his fists clenched at his sides. Neither Horton nor Rhoades bothered to ask Johnson why he had refused to work. The steward suggested to the foremen that, since Johnson had never been disciplined before, he could be given a pass home instead of a suspension. The request was denied, as Horton had expected it would be; Johnson was suspended and told to report to labor relations the next day. The UAW contract did not allow a worker to be fired off the floor, but every worker knew suspension at Eldon was tantamount to a firing. As far as Johnson was concerned, he'd just lost his job. A security guard, who had been waiting since the conference began, stepped forward to escort Johnson to the gate.

"I don't need no escort," said Johnson.

"I'll walk him out," said Horton. The foremen consented. John-son and Horton walked back through the plant, stopped by John-son's department to pick up his shirt, then started out of the building. As they neared the gate, Johnson broke away from the steward. "I'll be back," he said, and ran out of the plant.

An hour later he was back at the gate, dressed in light blue coveralls, a jacket tossed casually over his shoulder. Security guard Willie Pitts assumed Johnson was arriving late for work and made a routine call to the general foreman to report the arrival.

"Don't let him in," Rhoades told the guard. "He's been sus-pended. Try to get his badge if you can."

Pitts asked Johnson for the badge. Johnson was reluctant to give it up.

"You have to take it?" asked Johnson.

"Yes."

Johnson hesitated, then surrendered the badge and walked away. The guard assumed he left the plant grounds.

Inside the plant, general foreman Albert Volanski was alone in the production office. He sat for a few minutes drinking coffee and then made a call to the second-shift production clerk. As he hung up, the door flew open. Johnson burst into the office, pointed an M-1 carbine at Volanski. "Where's Jones?" he demanded. "Where's Jones?" Volanski stared speechless at the rifle as Johnson repeated insistently, "Where's Jones?"

Johnson scanned the office, then turned back toward the door. Volanski picked up the phone. Johnson swung around and aimed the gun at him again. Volanski dropped the phone. Johnson turned away, slammed the door aside and ran onto the plant floor toward Department 78.

He found Jones near the ovens and stalked him silently. Joseph Andrews on the No. 4 oven saw him coming. "Watch out," Andrews yelled, and Jones looked up, surprised, saw Johnson and ran. Jones ducked behind a skid box. Johnson ran toward him and the foreman ran again. Johnson aimed the M-1 at Jones and fired.

Andrews caught up with Johnson from behind. "No, man, that's wrong!" he yelled. "That's wrong, man!"

Johnson broke loose, pointed the M-1 at Andrews' chest, took two steps back, then turned and fired at Jones again as Jones tried to duck between the furnaces. Jones went down, tried to get up. Then Johnson stood over him with the rifle and fired again and again and again, hitting him four times. Jones lay still.

"Now I'm going to get Rhoades," said Johnson. He ran back toward the foremen's offices. He didn't find Rhoades, who was out of the building, but in the next few minutes he shot and killed another foreman and a fellow worker. Then, throwing his rifle—now empty—against the wall, he walked out of the plant. Outside, he felt the sun hit him and as it did he felt as if he were coming down, as if out of a nightmare in which he were a runaway machine that could not stop. When police arrived a few minutes later, he did not resist arrest.

In the auto shops throughout Detroit, James Johnson, Jr., was a hero. The frustrations Johnson took out in slaughter were universally understood in the plants by men and women with similar anger and frustration, similar fears. They knew instinctively that Johnson's actions were not "senseless" but born of a rage they themselves knew all too well. A Ford worker who later sat through Johnson's trial told a reporter: "Before I ever saw James Johnson—came down and sat through his trial—I knew the things that pushed this man into that. I realized, reading through the papers, that this man had not missed a day. He went to work religiously, not missing any time or being tardy, which for the average worker is very rare. So the only thing you can think in terms of this man was that he was pushed into it."

Dr. Clemens Fitzgerald, a Detroit psychiatrist who interviewed Johnson for psychiatric evaluation at the request of trial lawyers, explained it to me this way: "James Johnson came into the plant with a personality problem, with what I call a schizoid personality. That is a condition that bears a tendency to withdraw from reality; there's a tendency to daydream, a tendency to keep one's thoughts to oneself."

Johnson was well suited for the ordinary demands of his work, said Fitzgerald. "Out of his sickness he could withdraw into his own little world and do this monotonous, repetitive work. He was able to do this without difficulty when he wasn't harassed because he naturally retreated into his own little world. And to work on something like that you naturally have to become a part of the machinery."

But he was unable to withstand harassment. Being called "boy," being forced to work when he felt sick, being reprimanded unfairly—all of these harassments, said Fitzgerald, "put pressure on an ego that was already fragile, already finding it difficult to handle stresses of everyday living. It was hard for James Johnson to under-stand when he was asked to leave a job that he had worked up to, and go back to the pits, so to speak—the working in front of the fires.

"You see, for James, that was his world, because he wanted one day to have a house, to have a car; he had a lot of dreams that are normal, that, let's say, any individual dreams, in addition to other fantasies. To be told suddenly that he had to go back to a job that he associated with when he first started in the plant meant that he was being pushed out of the plant, and he couldn't take it. So Johnson, who wasn't a man who would drink or take drugs, utilized aggression. He lashed out at what he thought was attacking him."

Johnson was tried a year later for first degree murder in the killing of Hugh Jones. Although he could not remember any of the events following his suspension until the time he was arrested, he did not deny the shootings. His attorneys, Justin Ravitz and Kenneth Cockrel, stressed in questioning witnesses the serious problems at Eldon—unsafe conditions, an unresponsive union leadership, brutal management tactics, unequal advancement of whites over blacks, and increasing violence in the plant—which they contended put undue pressure on Johnson's state of mind. Coupling these conditions with Johnson's history of childhood poverty and mental symptoms, they had no trouble persuading a jury comprised principally of auto workers and auto workers' wives to agree that Johnson was "not guilty, by reason of insanity." He was committed to Ionia state mental hospital.

Judge Robert Colombo, signing the commitment order, recom-mended that the state "never consider releasing this individual back into society . . . This man is a killer, and this court feels would kill again if ever released into society." Hero or killer, all he'd wanted was the chance his parents never had, a chance for a decent life. When a reporter, visiting him in his cell while he awaited trial, showed him literature depicting him as a hero, Johnson replied, "I'm no hero. I never wanted to be a hero. All I wanted to do was to go to work, come home, and get my paycheck once a week."

Even before Johnson's acquittal, Ronald Glotta, a workmen's compensation attorney then associated with Johnson's criminal lawyers, had filed a claim for him, asking Chrysler to pay disability benefits and for treatment of his psychosis, which the lawyer argued was aggravated by stressful working conditions at Eldon Avenue. Strange as that idea might at first appear, there was strong legal precedent in a very similar case, which workmen's comp attorneys know as the Carter case.

James Carter was an hourly worker at Chevrolet Gear and Axle Plant in Detroit. He had worked without incident for three years on a nonproduction job. Then, in 1956, his trouble began. He was assigned to a job on the assembly line as a grinder. His task was to take up a hub assembly—a hub case and cover—place it on his work table, file the burrs from the holes that another worker had drilled in the metal, then replace the parts on the overhead conveyor line. From the start, Carter had trouble keeping up with the pace of the line. Soon he began taking two assemblies at a time, which was more efficient. But Carter's foreman, afraid that the parts would get mixed up, objected and ordered Carter to take the assemblies one at a time. Carter struggled with the job for several days, trying to keep up. Then, as he fell behind, he began furtively to take the assemblies two at a time. Again and again the foreman caught him and threatened to take him to the office if he didn't follow instructions.

Carter, who was becoming increasingly anxious, was admitted to the emergency room at the Highland Park General Hospital, where he was treated for "bronchial asthma and anxiety." His symptoms became so severe that his doctor advised him to take a leave of absence from the plant. Carter continued working, however, and a few days later called his steward and asked to be transferred to another job. The steward told him he'd do what he could, but meanwhile, he told Carter, he would have to finish out the shift. Carter returned to his bench. Later that evening his foreman again reprimanded him, this time for talking to another worker. Carter was immediately afterward overheard mumbling threats against the foreman, which were reported to management. Carter was called to the office and, when he attempted to draw a knife on the foreman, was fired.

Two days later he found himself in St. Clair Hospital in Detroit, unable to remember the events leading up to his firing or those of the two days between his firing and his admission to the hospital. During those two days his wife became alarmed by his unusual behavior. He bought appliances which the family had no need for—two cigarette lighters and a sewing machine—then drove his wife to his mother's house and kept the family up all night while he read from the Bible. At St. Clair his condition was diagnosed as "paranoid schizophrenia."

Donald Loria, a Detroit workmen's compensation attorney, filed a claim for Carter, contending that the worker should be compensated by Chevrolet because he had been disabled by a psychosis triggered by pressures arising out of his job. This was Loria's third such case. The first two were unsuccessful. They were similar cases, but were lost, Loria believes, because the employees began drinking, and drinking, he said, is considered "an inappropriate symptom to have for mental disease." Loria believed the referee found the Carter case more convincing because the man exhibited religious symptoms, "reading the Bible, praying, those things that are considered appropriate." Carter was awarded compensation in 1960 by the Michigan State Supreme Court, on appeal. It was the first such award in Michigan and a landmark decision.

A hearing referee found James Johnson's case similarly convincing and ruled in his favor in a decision in February 1973, citing the Carter case—"a case so close on the facts to the case herein as to be a very strongly persuasive if not controlling legal precedent." The referee rejected Chrysler's defense, which relied on legal nit-picking. A psychiatrist for Chrysler testified that Johnson's psychosis was aggravated solely by his suspension and that any other problems he may have had in the plant were ancient history and had no bearing on his state of mind the day he was suspended. In line with its own psychiatrist's testimony, Chrysler argued that the stress on Johnson came solely from his refusal to work on the ovens and his resulting suspension and that the stress therefore should not be compensable.

The referee described Chrysler's argument "as not being true to the facts herein." He also rejected Chrysler's contention that Johnson was no longer psychotic and disabled, noting that Johnson's psychiatrist, Dr. Fitzgerald, had testified that Johnson still had symptoms of paranoid schizophrenia and was regularly being treated with Melaral, a strong tranquilizing drug only used in the treatment of psychosis. The referee also pointed out that the Chrysler psychiatrist's inability to discover from his interview with Johnson any stressful events prior to his suspension "is probably well explained by [the psychiatrist's] own testimony that 'it was incredible that someone had filed a workmen's compensation petition on plaintiff's behalf.' That testimony," said the referee, "it seems to me, reflects accurately his attitude towards plaintiff and his problems which he may not have been able to conceal from plaintiff."

Workmen's compensation law, in Michigan as elsewhere, is built primarily on legal precedents. Earlier Michigan cases had established that a worker's disability need not come from a single, discernible accident or incident but could result from a continuing aggravation, causing disablement over a period of time, as often happens, for example, in back injuries. It was Ben Marcus, a Muskegon, Michigan, attorney, who won that change in the law in two cases known as the Shepard and Coombe cases. Before then, said Donald Loria, "lawyers were fighting over these very technical things: If you're doing heavy lifting over a period of time, you could say that back condition was an occupational disease [occupational disease became compensable in 1943]. But if it was one single lift that hurt your back and that wasn't accidental because you had intended to lift—in a single incident —then that wasn't an accident [and thus not compensable]. Lawyers were constructing these cases—'Now are you sure you didn't slip when you lifted this thing? Because that would be accidental.' And immediately when something would happen the insurance company would come out and take a statement: 'There was no slipping or falling.' And all it was was a lot of litigation over senseless things as to whether or not this was an accidental injury.

"But an occupational disease was compensable. Occupational diseases were generally considered to be silicosis, pneumoconioses, dermatitis, and so on. And then you would try to construct a back case and say it was an occupational disease. And you would have some doctors saying, well, of course every ruptured disc is just a process of gradual wearing away, from gradual stresses on the back. And some say, no, it's from a single trauma. You'd have all sorts of medical opinions. There was a tremendous amount of litigation which was good for lawyers and nobody else. Shepard and Coombe was the turning point."

The Shepard and Coombe cases, and others that followed, eventually established a principle in Michigan law, which is now generally accepted in most states. As the referee in the James Johnson, Jr., vs. Chrysler Corporation case explained it, the decisions "clearly established that an employer takes his employees when he hires them as he finds them, with . . . all their vices and faults as well as virtues, non-disabling weaknesses, defects, tendencies and proneness towards the development of disabling injuries and illnesses as well as their strengths and immunities from development of disabling injuries or illnesses. And if the demands, pressures and requirements of the total or whole work environment, when added to an employee's pre-existing . . . proneness or weakness . . . cause an employee to become disabled, the employer must pay compensation benefits to such employee."

Each worker comes into the plant with individual weaknesses. Some may have weak ankles; others, a tendency to heart or lung disease. And some may come with more fragile egos than others. Each worker has certain weaknesses because he is not a machine. And the theory of compensation that most courts have upheld is that the employer has an obligation to protect each worker and to keep him in essentially the same state of health as he was when he entered the plant.

So it was perfectly logical to compensate James Johnson, who, although since early childhood suffering mental problems, was able to work and function on at least a nominal basis—until pushed to the breaking point by stresses at Eldon Avenue.

AUTOMOBILES II

"It's Hell, It Really Is"

> You can't be treated like a boy at work and then, the
> minute you come home, be a good husband and a good
> father. It can't be done.
>
> *—Ford Rouge worker*

Most companies find, as Chrysler apparently does, that it is more
profitable to plan for disability and to compensate the sick and
maimed workers than it is to install safety measures to protect
them. It is hard to believe that companies can be so callous, so
brutal as to follow such a policy. Yet whether it is the brutality
of choice or inadvertence, indifference or ignorance, brutality is
what the evidence indicates. The daily stories are remarkable
for their similarities.

The story of Brian Flannigan, for example. Flannigan was a

142

young man working at the Ford Rouge foundry during the summer of 1969 to earn money to return in the fall to Notre Dame,
where he was enrolled in the School of Architecture. "I was working on a crankshaft table," said Flannigan. "The mold would come
down the conveyor in halves. My job was to pick it up and put it
on the conveyor which goes over our heads. The conveyors were
old and had a tendency to jerk and halt, which would cause the
dollies to wiggle and swing. Parts would fall off. It had happened
eight or ten times. We complained to everybody—to our committeeman on the shift, the millwright, the foreman. Our basic
complaint was there was nothing above our heads to protect us.

"That Friday night it emptied two molds, which came down on
top of me. Knocked me out. The guys on the table took me to the
medical office. Because I might have a back injury they sent me
to central medical. They took some X-rays. Basically they didn't
think anything was wrong. They gave me some Darvon because I
had a headache and sent me back over to the job. They gave me
a light duty slip. The foreman tried to put me back on the job.
I went and got my committeeman. He went to see the general
foreman. He let me go home only because he knew me—he knew
that I was hurt.

"I went home, woke up the next day and could not move.
Sunday was the same way. My mother drove me back to central
medical. For a two-week period they gave me Darvon. At no
time did they ever take cervical X-rays. They would poke me a
few times—had me come in every day for about two weeks and
then every third day. Finally Dr. Johnson said it was his position
it was psychosomatic. They sent me to Ford Hospital to have
two neurologists look at me because I was losing feeling on the
left side of my hand. Their position was, why wasn't Ford treating
me?

"I went to my family doctor and he sent me to Grace Hospital
for X-rays. He called me the next day, said 'Your neck is broken
in three places.' "

By this time Flannigan had been off work eight weeks. Because he had not been receiving compensation benefits, he contacted laywer Ron Glotta. Glotta was amazed to find that Ford was not only withholding payments, it had not even filed the accident form required by state law—in effect denying that the accident had occurred.

"Generally what would happen in a case like that," said Glotta, "the attorney makes more money the longer the case sits, because more benefits accrue, and you only get a fee out of the accrued benefits. But it was such a drastic case that I wrote a letter for a Rule Five hearing, which is that Ford Motor violated the act and is therefore subject to some kind of punishment . . . In a situation where a guy receives a definite injury on the job, they are required to provide a Form 100 to the state indicating an industrial injury. If they don't, that's a violation of the law. Ford Motor is notorious, particularly the Rouge plant. They never send in Form 100s. One of the guys out there, one of the attorneys for Ford Motor, said, 'If we send in Form 100s for every injury out there, we'd just be doing paperwork all the time.'"

They so seldom file, said Glotta, that "there probably aren't five filed with the state in the last ten years—other than, I suppose, in those cases where the guy dies there, or loses an arm, or something like that, you know, something so dramatic. A lot of workmen's compensation cases are that a guy's back gives out. And their position is, well, it could have given out anywhere. But Brian was standing there; a rather heavy item fell about ten feet and hit his back. He was clearly disabled; he was having serious problems; no medical care had been provided to him. No indication from the Ford Motor Company that they were liable for it. In fact, they said it was arthritis, or he was faking, or something.

"In order to start paying benefits you've got to send in a Form 100. But for all the other injuries, for which they dispute benefits anyway, they never sent in a Form 100. The reason for the Form 100 is then the state can explain to the man what his rights are and everything. And if they send in the Form 100, all sorts of problems have been solved for the worker's attorney. Number one, there's clear notice. Because the way a workmen's comp trial is, the guy comes in and says, 'I injured myself when I did this or that.'

" 'Did you tell anybody?'

" 'Yes, I went to the foreman and told him. He sent me to first aid, and they treated me, sent me back to work and I couldn't do the work and eventually I left and then I was fired.'

"That's the process of a comp case. Now if they sent in a Form 100, when he goes and says he injured himself by making a lift at work, then they've eliminated a whole legal issue that they always litigate: *notice*. Because when they send in the form, they indicate that he *told* them. But [without a Form 100] what happens in a trial, the worker says it and then the foreman comes in and says, 'No, I don't remember him ever telling me his back hurt.' Then the first aid person comes in and says, 'No, there's no note in here that he said his back hurt.'

"But you know, nobody down at the comp department really believes in that notice provision anyway. Because everybody knows the company is lying. And everybody knows that when a guy hurts his back, all the foremen know it, all of the first aid people know he hurt his back. So if the man says he told his foreman, everybody agrees that he probably did. Because you don't go walking around half-crippled and have the foreman just ignore it."

Again, the point of not sending in the Form 100s is economic. "If they were conscientious on the Form 100 situation," said Glotta, "it would cost them a lot of money, so they are purposely not anxious about that. And the national safety figures are compiled out of Michigan off the Form 100s, and the Form 100s are never sent in on any kind of regular basis." The company tries to get the injured worker on the group insurance plan, which may pay as well or often better than compensation. "They'll immediately pay group insurance," Glotta said. "They'll say, 'Here, take ninety dollars a week and don't fight the workmen's comp.' The guy figures he's going to get back to work anyway. And most guys do. After three or four weeks they go back to work. But a year later they have more trouble with their backs, they're off three or four weeks, and a year later they have more trouble. Each time it's been made non-industrial. Now at one point or another it's going to become so bad he can't work anymore. At that point, if each time a Form 100 had been filed and he'd been paid workmen's comp, he would immediately get his workmen's comp and he would get it for the rest of his life. Instead, what he has is litigation."

In the Flannigan case, Glotta forced Ford into filing the Form 100, and Flannigan began to collect his compensation benefits—sixty-nine dollars a week. "You can't survive on that much," Flannigan said. "I'm lucky I can live on sixty-nine dollars a week, but most people can't." Flannigan also filed for unemployment compensation but was denied benefits. "He's entitled to both, you know, theoretically," Glotta said. "Because the comp act says you're disabled from doing the work that you were doing at the time of the injury. Unemployment, you have to be ready and available for work, but only any work that you're able to do. But they beat him on the unemployment case, and the reason is clear. They don't want guys collecting both benefits. Because he has to stay on a sustenance level. If a guy were collecting both benefits, he'd be doing fairly well."

So that's the system. First, don't bother to keep the plant safe, because it's cheaper to pay off the crippled worker. Second, don't pay the man too much, on the theory that if there's money in it, he'd rather be crippled than work. It is a system based wholly on economics—in effect, on man's worth as a machine.

Zoltan Ferency, a Lansing compensation attorney, who was a former deputy director of the Michigan workmen's compensation administration and was a Democratic candidate for governor of

"The American worker everywhere is supporting society. The American society, they owe the worker something. Poor people come to work but they don't know what's happening. They are just trading their health away for profit."

—John Niewierowski, worker.

PHOTOGRAPHS BY THE AUTHOR

Robert Ferdinand—here with his son, Robbie—is permanently disabled and in great pain from exposure to a toxic ore while working at Kawecki Berylco, Hazleton, Pennsylvania.

"We try to keep the kids' spirits up. The only thing that bothers them is, their daddy's sick, and they want him to get better."

Leonard Bryant of Hanestown, North Carolina, worked in cotton mills from the age of thirteen, retired at sixty-two, and died a year later. Cotton dust causes lung disease, and Bryant had one of the dustiest jobs.

"It was right smart dusty in the carding room, on the job I was on. Anyway, it was pretty hard for me to breathe. Oh, it'll bother anybody. Finally it will give them a cough of some kind."

James Gleich, worker at the Anaconda plant in East Helena, Montana, where workers have suffered from exposure to lead dust and fumes and other chemicals.

"There's been several people that have been 'leaded.' This company is awful cheap. They don't want to do anything unless they make a profit. I hope to God my kids don't have to work here."

Chuck Forman (left), union president at the Anaconda aluminum plant in Hungry Horse, Montana.

"We've had men burned unrecognizable. Anaconda puts the money in production. The almighty dollar comes first. If they'd take the same amount of money they put into expansion and put it into better conditions, we'd have a real fine plant."

Juan Florez, farm worker in Tulare County, California, was poisoned when mixing pesticides.

> "Doctors [said the translator] told him they weren't very certain whether he was going to lose his eye completely, but he can't see very well with it. And his ear, he can't hear out of that. The specialist told him there's no hope for it to get any better."

Hugo Sanchez (right) was one of eighty California farm workers poisoned when pesticides were sprayed on the orchard in which they were working.

> "What happened was that the day that they were thinning the plums, they were spraying the trees also. Hugo, he came in, his nose was all stopped up. His throat, he could hardly talk. His stomach, and then all his face, had broken out with a rash."

At a nuclear plant under construction on the southern edge of Lake Michigan, there often are no guard rails, and heavy objects may fall on men working below. Said a safety inspector:

> "They killed a man while he was eating lunch."

Ronald Sayre, worker at the American Viscose plant in Nitro, West Virginia, was poisoned by fumes while handling raw rayon. Such fumes can cause terrible headaches and have psychological effects.

> "The company doctors diagnosed acute schizophrenic reaction. It really wasn't. It was carbon disulfide poisoning. It was making me think weird thoughts."

The Syracuse Foundry, Syracuse, New York. Foundry workers often must labor in a thick haze of dust and acrid fumes—and risk silicosis, cancer, and other serious afflictions.

"It's a terrible place to work, and they don't do nothing about it. They laugh at everybody working here."

A welder at work. Ken Bellet, a welder at the Bethlehem Steel plant in Buffalo, New York, said:

> "What happens, when welders are welding, they are breathing the dust and the fumes. I haven't welded any day where my nostrils aren't just plugged at the end of the day, to where I can hardly breathe."

Above, Kellogg, Idaho, site of the Sunshine Mining Company, scene of a 1972 silver-mine disaster in which ninety-one men were killed. At right, miners' relatives grieve at news of death.

"It hurts, it hurts. I don't know if we'll ever get over it."

Ira Sliger (above), hoistman for Sunshine, suffers from what he calls "dust on the lungs"; he has lost a lung.

"I look big enough to eat oats and pull a plow, but if there's any exertion in it, I can't do it. I've put a lifetime in these mines. I think I've put in pretty near enough."

Tom Wilkinson and Ron Flory, the only two miners rescued—living reminders of the ninety-one others who died under the ground.

The grave of a Sunshine miner, one of the many American workers—
a hundred thousand a year—who are killed by their jobs.

Michigan, pointed out that the basic premise of the law is false, because people want and need to work. "The typical thing that happens," he told me, "is, the guy goes to work, or the girl goes to work, that day and the trauma happens. It's only then that the whole impact hits—the loss of job, loss of well-being. Hits them all at once. He has to go home and explain to his wife and kids, 'I am no longer able to work for a living.' Unless you're in this business, you don't understand what a feeling it is to be a worker and not be working. Until it happens, even the man doesn't understand it—the Puritan ethic. I've had cases where a man loses his potency, his manhood. You become a useless person."

These are strains on the person that can be expected in the best of circumstances even if the company immediately and voluntarily begins to pay him compensation. But often simply trying to get that compensation becomes a major problem, dragging on in many cases for years, focusing the man's attention morbidly on his disability. During this time he is at the mercy of his lawyers, who may or may not have his best interests at heart, and of his employer or the employer's insurance company, which may be withholding his payments in an attempt to pressure him into settlement or returning to work.

In Detroit there are law firms that commonly convince their clients to press for settlement rather than proceed with litigation, regardless of the merits of the case, because the lawyer makes more money that way. "As soon as you get volume," Ferency said, "sheer economics tells you that settling a thousand cases at fifteen percent is more than litigating two hundred and fifty at thirty percent. Lawyers will deal off one case against another."

In the Detroit compensation courts, it is a common sight to see a docket listing fifteen or twenty cases, all from Chrysler or Ford, all represented by one attorney, who will attempt to settle all the cases in one session, bargaining with company lawyers, trading a thousand dollars off Mrs. Hugh's amputated leg for an extra thousand on Mr. Frye's broken back. And their clients, grateful for the few thousand lump sum settlement—probably more money than they've ever seen at one time before—may never know that they may have been entitled to twice or three or four times that amount, or more.

The Johnson, Carter, and Flannigan stories are varieties of a tale that is replayed constantly in the auto shops. It is a story of basic conflict between the worker and his work, causing the countable injuries—the amputated fingers, broken bones, back injuries, and such—but also the more hidden but equally devastating effects on workers who may never appear in the workmen's compensation docket.

Dr. Clemens Fitzgerald, who had studied hundreds of auto workers, believed these effects were pervasive, and would prove to be so far beyond what anyone had yet chronicled. "It takes a certain type of person to work in the plant," he said. "This is why some people turn to drugs and alcohol, because they think this is what reinforces them and enables them to do it. For other people—those who are very weak—they develop some type of syndrome that we're not even aware of, because how many people really study the poor worker in the plant? I know that very few people have studied the black or Mexican-American worker because no one is really that interested in them, unless they end up in a state hospital or end up in jail for some offense."

One of the biggest stresses in the auto plants, Fitzgerald said, is caused by treatment workers receive at the hands of their bosses. Just as it affected Johnson and Carter, it affects the average worker. Especially, Fitzgerald found, it affected the black workers with whom his studies were primarily concerned. "They were treated as if they were boys—they were called that, in many cases. They were treated as if they had no sense, no intelligence, nothing to offer the plant but their hard labor."

A foreman has a great deal of power, which he can use arbitrarily to reward his favorites, or, too commonly, to punish those

he dislikes, even to the point of causing actual injury and permanent disability. Perry Rickman was one of many victimized by the foreman's all-powerful command.

Rickman's troubles started in the early spring of 1965 when he slipped and fell in the company parking lot after a freezing rain, injuring his back. He was working at the Ford Lavonia Transmission Plant in Detroit, and a Ford doctor, he said, told him, "There's nothing wrong with your back, Mr. Rickman. You just got too much belly." Nevertheless, he was put on a light-duty job.

He was on light duty for a week, then returned to his regular job as a mill operator. "I was milling transmission cases," Rickman told me. "The pig iron, the rough pig, when it comes from the foundry, weighs about sixty-two pounds. They would be in a basket. I would bend over and take these out of the basket. I would fit these in a machine and I would clamp it. The machine would revolve and wheels would mill the bottom and the top. While it was milling the bottoms and the top, the machine would revolve and another transmission case would fall out. I'd have to unclamp it and take it and put it on the conveyor line. On the average I would do anywhere from six to eight hundred a night."

Back on his regular job he began having back pains. Back pain in industry is so common as to be prosaic, but is a serious threat to livelihood, and Rickman took it seriously. He spoke to his foreman, who told him if his back was hurting he didn't have a job for him. He was transferred to another job—as a bore operator, working with lightweight parts. He could handle the job without any problem. After six months in that department, he was upgraded to an inspection job which involved stacking parts. Again he began to have back pain. "I went to first aid, and they gave me another of these elastic belts to wear," Rickman said. He continued working but asked for a transfer to a lighter job. He talked to his steward, too. "I asked why didn't they give me a lighter duty job to do. He said they didn't have one. The steward told me—the only

thing I could do was do a fair day's work for a fair day's pay."
Rickman asked the steward to file a grievance for him, but the
steward refused.

"I was missing quite a few days from work because of the back
problem. And they asked me why I was taking off, and I told them
I didn't feel well. One particular day I had a flat tire, and after I
had finished changing the tire my back began to give me a lot of
problems. Some fellows that worked with me passed by me, and
they saw me and asked if they could help me. I said no, but if they
saw the foreman to tell him that I had a flat tire and I would be in
as soon as I had the tire fixed. I had it fixed at the time, but I just
didn't feel like going in, and I went all the way back home. I went
back the next day and the foreman—his name was Fred Seaborn—
he wrote a grievance on me for excessive absenteeism, and we went
up into the office and had a meeting. The committeeman [steward]
came in and he asked Fred Seaborn, he said, 'Well, Fred, why don't
you give the guy a break and let him alone? He's one of your best
workers, he just misses this time, and he says that he's having these
problems.' Fred said that he couldn't, and so the committeeman, he
told me, he says, 'Rickman, why don't you buy Mr. Seaborn here
a fifth of Canadian Club and maybe he'll drop this.' And I told him
I wasn't buying him nothing. And they wrote the grievance, and
they docked me, I think it was half an hour."

Soon after, he was transferred, and his back problems continued.
Eventually he was upgraded again, and assigned to a job balancing
flywheels. "Each balancer would do five hundred pieces a night.
But I had no back trouble. It was the high overtime department
in the plant. We were working seven days and the balancers were
working twelve to sixteen hours. I would get off in the morning
eight or eight-thirty, and I would be back that evening at eight.
Sometimes I wouldn't get off until noon and had to be back at
eight o'clock that night. And I got sick. They sent me home
from the plant. I went to see the plant doctor because I felt dizzy,
and I had elevated blood pressure and they told me to see my

doctor. I saw him and my blood pressure was sky high. He gave me some medication to take, but the blood pressure didn't go down. And he noticed that my hands were perspiring and all down my back on a normal day, and he says he thought I had some type of a nervous disorder. He wanted me to see a psychiatrist. The psychiatrist examined me and they came to the conclusion that it was the high overtime that I was working. And they put me in the hospital."

Two months later he went back. He was returned to the balancing job, but his machine had been moved so that he had to lean over the conveyor in order to reach his parts. He began to have back pains again. "Then we had a time study, and they upped production from 500 parts to 1,182 parts per person per shift. Shortly after that, my back problems, they really increased."

Rickman tried to make production but couldn't. "The best I ever got, I got 800 parts one night. Shortly after that, they began to cut the department. They took two balancers off each shift where we had seven men balancing. That made only five. And I had to leave the midnight shift and go on afternoons, working for a new foreman. I had problems with this man when I was working as inspector because I caught him stealing parts. Parts had been put in a basket, and they had been okayed. So to get him a little larger production, someone had pulled the tag. I don't know—he says he didn't do it, but I'm almost certain he did because no one had been over there but him.

"Well, he jumped me right away, told me I was a smart-ass and we had had problems before, but I was working for him now and I was going to do what he said to do. We worked a straight eight hours. We would start at four and get off at twelve. We originally was supposed to get only twenty minutes for lunch hour, and a lot of the men was taking forty-five minutes or an hour for lunch. But he insisted on me being on the job. I couldn't take but twenty minutes for lunch. We rocked around for a long time and one day I came in, this was March of '69, I came in to work and we had

had some parts that had been sent back from some place in Ohio, and all these parts were out of balance.

"The foreman told me, he says, 'Rickman, we got so many C's'—that's the part number—'we got so many C's that came back from some place in Ohio and they're out of balance.' He said, 'I want you to get your ass over there on the balancer, I want you to set it up for those parts—it's a little better than a thousand—and you're going to get them today, every one of them.'

"When the whistle blew for us to start work, I went over and I set my machine up. He had the hi-lo driver bring two four-by-eight baskets of parts. It was 500 to 550 pieces in each basket. I had to bend over the basket, take the parts out, stack them on the table and then run them. I run 300, 350, I forget, and the conveyor was full so I had to take the parts off the conveyor and stack them in an empty container. And I was doing actually double job. I was doing the balancer's job and the production checker's job. I had sent for my committeeman but he hadn't got there yet.

"My foreman came down and asked me how was I getting along and I didn't say anything to him and he kept badgering me and running his mouth and we had some words, and I told him I wasn't going to take the parts out of the basket, stack them, run them, and then restack them. He told me I was refusing to work, and one thing led to another, and I told him to get my committeeman down there and get him down there immediately. He told me that he would, and he walked away.

"He came back with a fellow by the name of Gibson. He called him 'The Hippie,' I don't know his first name. And Gibson would take the parts out of the basket and stack them for me. And then when he was finished, he would sit down behind me on a piece of cardboard and he would wait until I had 300, 350 parts, and he would take and stack them. Him and the foreman sat behind me and watched me run those parts and had just a bull session, you know, running their mouth, laughing and talking—and I don't know, I'm just saying this, but I felt they were laughing at me because I was working so hard.

"I did the parts. It was better than a thousand. I worked right up until ten minutes to twelve. All the men were standing back there, they had quit work approximately half an hour before, and all the men were standing back there watching me work, laughing. When I finished up, the foreman said, 'Now you're finished, Rickman?' He says, 'Come on, I'll buy you a cup of coffee.' And I told him to go to hell. I still hadn't got a committeeman. I was having back pains—and that was the first time I started having the pains down my left leg. All the toes on my left foot was numb."

From then on, he was in constant pain. The foreman put him on another machine. "When I run out of parts on that machine, another man had gone to lunch—his name was Glendenning, he was the foreman's boy—the foreman made me go down and run Glendenning's job. I asked for a slip to go to first aid. I told the foreman my back was hurting me, he told me I was full of shit, to get back on the job. I refused to work and went to first aid. They sent me home. When I came back to work, he asked me if I had a slip from the doctor. I told him no. He told me, 'Well, your ass has had it.' I asked him what did he mean. He said, 'You'll find out after a while.'

"They had an air make-up unit on the column there, and this air make-up unit—it was still quite cold outside—and this air make-up unit was blowing cold air. I was hurting all up in my back and my shoulders and all in my hips and down my left leg, and he told me that was tough. I told him let me get a slip to go to first aid. He said no, that he was going to write me up. I told him to get my committeeman. He told me, he said, 'I'm not getting you any damn body.' I told him, 'If you're going to write me up for not bringing in a doctor's slip, I hope you do, because I say I'm writing you up.' The foreman told me, he said, 'That's good.' He went up to the office and he came back with a Form 28. And he sat down and he signed it and he handed it to me. He said, 'Now you don't work in this department anymore. You put in an upgrade slip, I saw it in the office. Now if you can get yourself a job in another department,' he said, 'that's good.'

"I took the upgrade slip over to another department, and a man by the name of Whit had sent for me approximately a month before to come to his department, but my foreman had refused to release me. And Whit accepted me. I guess I worked approximately a month or five weeks. And when I just couldn't take the leg pain anymore, I took off and went to the doctor. I couldn't put the left foot down at all. I had to hop and I had a cane I was using."

Eventually Rickman was examined by a neurosurgeon, who diagnosed a sprained and fractured lumbar vertebra. Rickman spent six weeks in traction. Then he went back to work. "Ford's given me a restriction of no bending, no lifting, no climbing steps, ground level work only, day shift work only. The employment office told me that if I could go back into the department and do the job that I was doing before all this, they had a job for me; otherwise they didn't have a job. I talked to my committeeman. He said there was nothing he could do about it. He said that's where Ford has the advantage over you. They put all these restrictions on you, and if you can't do your old job, they send you to another department. The foreman in the other department says, 'Well, I need a well man, I don't need a sick man,' and they won't accept you."

Shortly after Rickman went back, his leg buckled under him and he fell on the stairs at home, breaking his hand. He was off another five weeks. When he went back he fell again at work. The foreman sent him to the plant doctor. The doctor sent him home. "He said there was nothing they could do, they wasn't going to let me work anymore. He said the best thing for me to do was to see my family doctor and, if I could get it, to try to get workmen's compensation."

Rickman's sister worked for lawyer Ron Glotta, and he asked Glotta to take his case. He won a compensation award two years later, shortly after I talked to him, for total disability. At Ford he had been making $177.44 weekly; with compensation benefits he received $86 a week. I asked him what he thought about James Johnson, the auto worker who killed his foreman. "The only thing I know I read in the paper. But I worked in the factory myself, and it's hell working in the factory. I don't know this James Johnson personally—only what I read in the papers—and I'm inclined to believe by the experience that I've had myself that it may have been something that made him do what he did. Working in a factory is hell. Some foremen treat some guys real nice and then the same foreman, he'll treat other guys like dirt."

The pressures of the plant can cause injury to workers in even more subtle ways. Harassment, production pressures, long hours, poor maintenance, lack of safety offer workers little choice. They must find some way to cope, short of leaving the job—usually an unacceptable alternative. And most of the attempts workers make to deal with the plant environment are injurious. Workers often talk about how they develop a sort of split personality in order to endure their work. As one man puts it, "My body would work mechanically, and in my mind I'd be off someplace fishing, or I would be off with this best girl I had just met, or anything like this, automatically; you could completely divorce yourself from what you were doing here and you'd go someplace else."

Clemens Fitzgerald, the Detroit psychiatrist who has studied auto workers, said this sort of daydreaming is necessary because the alternatives would be too destructive. "For many people, to be aware of what is really going on, they'd become so angry that they would either become very aggressive and destructive or they would become so depressed that they would be immobilized. So in order to do their work, they just have to have another personality, where they are really working, but at the same time they're off in a dream land or a fantasy land—an ultra-state—and frequently this can be brought about by the use of alcohol or drugs."

But the daydreaming is also unhealthy. "It's unhealthy," said

Fitzgerald, "because anytime you have ultra-states, you put more pressure on the ego to maintain these ultra-states, and to bring you back to this ultra-state once you leave it. The result is that you have people who have breakdowns at one time or another; you have alcoholics, you have drug abusers or drug addicts, or you have people who go out and commit crimes."

A similar mechanism acts on workers who appear to be uninterested in safety, who won't wear hard hats or safety glasses, or who take unnecessary chances. Fitzgerald said they behave in such ways because "they have to deny these dangers in order to stay in there. Frequently workers use 'defensive denial.' They say to themselves, 'This injury is not going to happen at all.' Then they can maintain themselves. The only way they can say that this is not going to happen is by acting like it isn't going to happen. So I don't wear my hard hat, because that implies there's something up there that might fall on me. So I act just as if nothing was there. Of course this is bad because denial is a primitive defense. Because they're denying what is real. Their touch with reality is slipping.

"Then you have another kind of worker that is rebellious. He is rebellious because he sees the company as a hostile establishment that really doesn't have the interest of the worker at heart. These workers say to themselves, 'The company doesn't care about my safety, I don't care about the company.' Many of them have said they aren't concerned about doing good work. They've seen a buddy lose an arm, or another buddy blinded in the eye, or another one having dizzy spells from being hit over the head and nothing being done—a lot of commotion, a lot of people investigating; then after a few months, things return to the normal status at the plant, which is a condition that is less than desirable, both from safety standards and from an emotional climate."

The effect of plant pressures is also seen in workers' emotional reactions outside the plant. "One of the things that intrigued me," Fitzgerald said, "was the number of fights in the homes of factory workers that I studied. After talking with them I saw that a lot of their anger and hostility, directed at the foreman or company itself, was displaced toward their wife or their children, so that we saw excessive punishment of the child."

Workers know this—at least they can see it in fellow workers. "These are good people who work in those plants," a black worker from Ford Rouge told me. "They work hard, much harder than they should, but it isn't the hard work that causes the problems. I wish I could take you into my plant and let you see guys sixty years old handling forty tons of steel and laughing about it. They may go home beat, but they're proud men, they have a reserve and a humor that allow them to perform this hard work and still smile and still joke. These are the older guys. The kids are different. The kids come in angry and leave angry. Some of the older guys do too. But this is what causes the social ramifications that we have out here in the street. The murders, the broken families, the drunk drivers. You can walk out in the parking lot out there and see it. Brand-new car today and two weeks later it looks like junk. He was drinking and driving. He hit a telephone pole. He busted someone else's car. He hit someone backing. At change of shifts at Rouge on the Schaefer Roadside, it makes Indianapolis look like a cake walk. And you say to yourself, what in the hell makes them drive like that and you realize—just trying to get away from that place. And we've had guys killed half a mile from that plant before they got to the freeway, because they were driving sixty and seventy miles an hour when they should have been driving twenty-five or thirty, but they just had to get away from there. And they drive like crazy. They are crazy. They've been driven crazy by working conditions.

"And when I say working conditions, as I understand it, if my memory serves me correctly, the top temperature on the Sahara desert is 137 degrees. It's over ninety outside today. In my plant, if it's over ninety outside, on the job I had until March it will be 140. And guys will be handling from twenty to sixty tons of steel in eight hours, with a thirty-minute break before and after lunch. They will work soaking wet. The welders will have to have someone peel their soaking-wet coveralls off them. The guys handling the steel, if it happens to be rusty, they're just—oh, you can just imagine the combination between rust and sweat and flesh.

"And then they go out of there. How are they going? What's their reaction? They go home and the old lady says, 'Cut the grass.' The kids say, 'Play ball with me.' And you'll hear a woman say, 'You no-good nigger'—you know how it's said. They simply don't understand."

The worker believes he's being wronged not only by the company but by his own union as well. "You see," said the Ford worker, "you have to understand that the UAW is a bureaucracy, no different today than the company bureaucracy or a government bureaucracy. They are powerful and they are rich, and they reward their friends and punish their enemies. You have to remember that a UAW officer is on leave of absence from whatever his job was in the plant, and if and when he loses an election, or is replaced on his job, where is he going? International reps make a minimum of $15,400 a year. They're not literate people, they're not salaried. But they know how to follow orders, and they know how to con guys, and they're good salesmen. That's how they got the position. But what guy making fifteen-four, plus expenses, is going to make noise and rock the boat and get kicked back to the job where he was making nine thousand and was working his fanny off? You know, most human beings aren't going to go that route, so they go along with the program. In the process of these guys following orders and making it for themselves, they have climbed over the dead, injured, maimed, schizophrenic, wounded co-workers who they were supposed to represent. And that guy who didn't get represented is there everyday and is like the kids at Eldon Avenue, or my plant or any other. He knows he's getting shafted. He knows that both his company and his union are doing it. He's angry about it. He responds numerous ways. He's out on the street on Friday and Saturday, he cuts and shoots. At home he has broken homes. He's not a decent father or a decent husband. And it's—the pressure is just unbelievable. I'm amazed they survive. But it's a tribute to those poor guys that they do."

A woman who has worked in several auto plants, most recently in Chrysler's Vernor Trim Plant in Detroit, told me that while men's reaction to plant pressures is often seen in violence, women show different responses. "There are a lot of screaming-fit type things," she says. "Crying. A lot of alcoholism. At least it's more noticeable than among the guys. There's a lot of drinking on the job. And a lot of narcotics on the job, very heavy. Very easily accessible. A lot of that stuff. Among the women, there are a few that get carried out in the course of a year, dead drunk, and I myself, who don't drink very much and can't take very much liquor, spend most of my lunchtime getting a drink or two down me to get back. It depends a lot on your foreman. One job particularly, the floorlady was such a bitch, you know, in every sense of the word, it was just horrible. In order to keep from hitting the woman everybody had to have a drink at lunchtime to make it back. There's no other way around it."

This woman, a thirty-six-year-old white worker who asked that her name not be used, described situations that make alcoholism and drug use understandable.

At Vernor Trim, where she worked as a checker, most of the work force was female. Their job was to take huge rolls of vinyl fabric and cut and sew it into auto upholstery. "I was working at sewing front-seat backs," she said. "Somebody else might be working on front seats. Now, every model change, no matter how minute, there are new production standards that are set. And the company says that you are supposed to do, say, twelve backs or seats an hour, or in some cases a hundred an hour—it depends on which operation you are on. These are estimated rates of work—and they have very little to do with the real ones because nobody knows, at first, how long each job will really take. On some jobs—

oh, like on small cars—you don't have that many inches of sewing.
Or finishing welts—the welt which is wrapped around the edges;
after this welt is put in, in order to keep the vinyl from buckling,
there is a finishing. You turn the inside in, and then you have to
sew right close to the edge. Well, if you got very stiff and very
thick and heavy vinyl, and if you're sewing on a curve, it's not
the same number of stitches per minute you can sew if you got
a seam going up and down.

"So it's really not related, the job to the quota, and yet the
girls are expected to do this. And you're browbeaten during that
first month after there is a model change—you have to make that
quota. If you can stick together and avoid the pressure and all of
you decide how many you are going to produce each day and
let it ride at that, you can usually break it. Of course, that's the
hardest thing to achieve because there are all sorts of mechanisms
to break up that solidarity, from shifting people around if the girls
are too tight—the next thing you know, that one is off somewhere
else. The women fight it. They have their friends and they like to
work near their friends, especially if they have been in a plant
ten, fifteen, twenty years.

"For another thing, you're used to a certain operation. You have
certain muscles. You might have to pull on one side, push on the
inside of the other. Or sometimes people are shifted just hap-
hazardly. There were women who as long as vinyl rooftops have
been made have been doing vinyl rooftops. Then the company
started doing the vinyl rooftops somewhere else, up at Detroit. So
they shifted the women to other jobs. They put them on French
sewing for a French seam, which is a bitch of a job. One woman
I know with twelve years' seniority, the job that she had for years
and years and years was rooftop. Now she's been put on French
seaming. Any slight misalignment and you get caught into the little
seam itself, and you can't get the machine back out and you can't
repair it, because once the machine goes into the vinyl it makes
holes and you can't sell a seat cover with holes in it. And you know,

if there's a lot of repairs, all hell breaks loose. You know, they don't make money on it.

"They're supposed to shift people around according to what they can do, but in this case, when the company moved the rooftops over to Detroit, they had four girls who couldn't produce, couldn't make the production quota on these other complicated jobs. After one week, they started bugging them. After two weeks, all four of them have gotten foreman's reports. Which is a warning. The first time you get a verbal report. The second time you get a written report. The third time you can be laid off a day. The next time, three days, or five days, ten days, thirty days, you're out. It's an automatic disciplinary procedure that leads up to firing. Chrysler's own "thirty-and-out." Another aspect of their thirty-and-out is what happens to the older workers when they get shifted to harder jobs. They are either forced to quit or take an earlier retirement. Which means they get very little money."

Women at Vernor are often afraid to file compensation claims, she said. "The general feeling is that if you're going to do that, you're going to lose your job. And there's the attitude that they will never get another job." Many are even afraid of losing their jobs if they stay home when they are sick. "One thing that happens—I haven't particularly heard of it at Vernor, but at other plants I have heard about it—where the foreman just plain refuses to accept the doctor's excuse. Like there is this one incident, this is at Mack [Mack Avenue Stamping Plant]. The guy had a hundred and five fever. First aid sent him home. They called his home two days later, because when you're out three days, your insurance starts. So rather than pay insurance, they make sure you're back at work. They called his home and told him that he had to come back. And he came back the third day with his doctor's excuse and the company refused to accept it.

"And then in addition to not accepting the doctor's excuses they call your home and send somebody out to verify you're sick. And if you're sort of able to walk around, you're back on the job.

Or at least they threaten you with loss of job. There's not much you can do. Because they can always put you on a harder job. They can always up your production, there are all kinds of things they can do to make it completely unbearable.

"One of the things against everybody, including a lot of the younger women, is the paternalism, the degrading attitudes that are presented, like anything that happens to you, no matter what it is, is of course your fault. And you're patted on the head and told, 'Now, now, if you just think better about it, I'm sure it will all be all right.' You know, like you're a little kid. And some of the women get mad, but some of the women in a sense succumb to it and are even more babies. There is just this constant personality conflict, and it's enraging, absolutely enraging. And there's absolutely nothing you can do about it. Not just to me —it's enraging to half the people there. I've seen the women foremen come out of the office in tears and hysteria, because they obviously treat them the same way too. It's not just the man-woman relationship, because the foremen in other plants get treated by supervisors in the same fashion. Many of them quit. Many of them ask to go back to the line, preferring that to staying foreman."

People often avoid using safety equipment, she said, because "with most of the safety equipment you can't keep the speed up if you're going to use it. So I have heard many guys say that the only time that they use the safety tongs instead of their fingers is when the safety man is coming around, because they'll get written up if they don't use them. And then their production will fall behind, and soon as he's gone they will throw them away to get caught up, because the hassle you have to go through when you fall behind is ten times worse than the possibility, I suppose. Nobody ever thinks that it is going to happen to them—that they're going to lose a hand. Naturally, nobody ever thinks that. They're going to be careful, and they really don't need the tongs.

The crap they have to put up with about it is too much to handle. It's that kind of a problem.

"At Vernor, there were dust particles in the air that I think a lot of us were allergic to. It's a kind of a plastic, it's the foam stuff that goes behind the seats, behind the vinyl, and what it is, the vinyl itself has a kind of chemical backing in it. Then there is a foam, and then there is a backing on the foam which is some kind of stiff cheesecloth or something. And the stuff is very irritating. Just to handle it, you end up with little bumps all over you. I don't know if it's just some people, not everybody, just some that might be allergic to it—it's hard to say. And then the dust, the lint from it is all over, and everybody has colds most of the time, and everybody attributes this to the fact that there is all of this lint around. It's probably true. As well as that the place is drafty as hell. And rashes were pretty common. Everybody attributed it to nerves. If it got real bad you would go down to first aid and they would swab it with alcohol.

"They used to have floaters around a number of plants, which meant that if you had people out sick or someone had to go down to first aid, you had somebody who sort of knew most of the jobs and could fill in. A lot of that's been done away with. They've been given permanent jobs of different kinds. So you have fewer and fewer floaters. What that means, of course, is that in order for the foreman to let somebody go to first aid, he's got to wait a few hours till he finds somebody that might be a floater or a relief man to relieve the man. Only if it's dreadful, if he's passed out or something, they got to do something. But it takes almost that."

Steward representation at Vernor she described as "virtually non-existent."

"Most of the time you can't find them. Let me give you an example. I was told to rip out a seam for a whole stack of repairs that had to be done. Now, that's supposed to be a sewer's or repair job, which pays more than I was getting and I'm just not supposed to do it. I sort of didn't think I was supposed to do it and I sort of muttered something to the floorlady, but this was one who was a bitch, and who you could barely get away with looking at her cross-eyed. So she started riding me up one side and down the other, and I started ripping seams. So the steward comes around with the assistant committeewoman and the committeeman. And just gives me hell. I didn't say word one. And then one of the other girls got so mad, I guess because I wasn't defending myself, she jumped up and left her machine and said, 'What do you want from her? She's a good checker. She's only been here a month,' et cetera, et cetera. 'Oh, you only been here a month? Well, you should know, we won't be harsh with you. You just treat this as a lesson now.' I said, 'Well, what am I supposed to do if the foreman tells me again to do a job I'm not supposed to?'

" 'Don't do it! Don't worry about it! Don't worry, we'll protect you. You don't have to do it.'

"So of course, after they left, they never said a word to my foreman. They never said a word to her and they never do. So I asked some of the girls around, you know, what will happen. They said keep your mouth shut and do whatever she tells you because if you don't you will be out of here on your ass.

"On paper, you can really live the beautiful idyllic life that Reuther had in mind. But there's nobody to enforce it. For example, the union is supposed to check out the speed of a line on the assembly line at any time. But first of all, there has to be somebody from the union willing to fight the company to do it, which there isn't. And then there has to be signed complaints from the people on the line—and they're not going to do it, because whether it's right or wrong, they're going to have to eat a lot of shit after that."

Much of the trouble is with the grievance procedure, which works so poorly it discourages workers from attempting to use it to solve their problems. "When you figure that at Eldon, when the yearly memorandum contract came up, there were five hundred-some grievances out of which three or four were settled," the woman said. "No one knows what happened to the rest. And the basic bad thing about it is that you're guilty until proven innocent. For example, you say that the company has violated job specifications and they're trying to force you to do so many number of things and work at a certain pace when the agreement calls for such and such a pace. Well, you have got to work at the company's pace until it is proven otherwise. And if you can't do it, then you have to have the stamina to fight it.

"For example, if you're behind all the time, it's much more work for your fellow workers around you. Everybody's on your back if you don't do it. You just don't want to be a ninny and say, 'I can't do it.' Because generally speaking you can. It might take ten years off your life but you can do it. Maybe you go home and you can't do a damn thing else, but it's physically possible for that given day or the next given day to do just about anything that they ask you to do. And if there's any kind of weakness anywhere —if you have a tendency toward a hernia, or a who-knows-what, you're bound to get it. So each person tries to work at their own pace, of course, but then the other guys around you don't let you, because it adds work to them, and if the whole line gets in the hole, all hell breaks loose.

"And you have to have the stamina to take that, to argue with the foreman, to speak up for yourself, to fight it through the channels, to know the channels, to know what to say legally that won't get you fired. You can never say, 'I won't do it.' You got to act like a little kid all the time and say, 'Well, I'm trying as hard as I can'—practically bursting into tears if you're a woman—or get furious, mad, go through the routine. Not everybody has the guts for that. And I don't mean guts in the sense of being strong, just the distastefulness of going through that stuff. Not everybody can do it. And if you can't, you break your back or die ten years younger. So I am sure there is a great number of people who will bitch, but who will not file a grievance. Because once your name is down on a grievance, you're on somebody's shit list after that. They're checking to make sure quality control is there to make sure you aren't making a mistake. The time study man is making sure that you are working every second. Et cetera.

"The union has the attitude—this is what they say in public every time they come down to a local meeting to squash a revolt or something: 'You guys yourself, you got all the power, this is the final body here, you just got to get yourself together and do it.'

"The union, do-gooders of all kinds, are always saying that— 'Yah, if you people would just organize, get yourself together and do something.' Well, hell, everybody knows that. How? When? Where? What do you do—how do you do it so you don't get fired? How do you do it so you don't have to put up a fight twenty-four hours a day, on the job all the time? Now that's the problem. Everybody knows it needs to be done.

"Reminds me of one guy, working up at Chrysler Truck, one summer they had students working there and one student, I guess a radical student, was really giving them a big rap. These two guys—a black and an older white guy who was in on the early union organizing days—they were talking about it, having a good time. This young kid was listening, getting a few words in here and there. At some point, when the old guy was getting sort of whimsical and the young kid's getting impatient, he said, 'What I don't understand is, well, don't you guys know you're *exploited?*' The old guy sort of sits back, you know, and says, 'So what else is new?' You know, like somehow just this knowledge will free you. Like if you know you're exploited, therefore you're no longer exploited. Or if you know you've got to get yourself together, therefore you're together.

"There's something wrong in that attitude, some lack of under-standing of the problems and the process. And most of the workers get that attitude from everybody, the union as well. Like the union says, 'Well, if you really think the union is rotten, well, just get yourself together and vote us out.' Well, we'd have done that years ago if we knew how. But there's all the slander campaigns and the undercurrents and the deals and what have you, and you don't know what to believe or what the score is. Because you're just an individual. You're not together. Nobody has won enough confidence to be able to say, 'This is so,' and have people agree.

"And then the whole bureaucratic apparatus is such as to avoid any really getting together and doing anything. In many ways the UAW isn't corrupt—I mean corrupt in the traditional sense of money-grafting and what have you—they're not at all. They're very much socially conscious. But they talk a lot about wanting to have changes. And they don't go seeking out where those things have to be done at. They don't have the legmen to do the work. With safety and health, like almost in everything else, they're all for it, it's just that there is nobody to carry it out. There always is a lot of talk about it in contract negotiations, but there's no change in the contract itself. Locals are supposed to have a safety committee and safety problems and grievances are supposed to come up before the safety committee. Of course, unless there is a local that's relatively militant, or at least even reasonable, no grievances go anywhere, and safety is the least of them because the UAW has trained its people for so long to think in purely monetary terms.

"For example, I was talking with the guys in my local last night. It's an amalgamated local. I work at one plant, they work at another. They work over at Mack. About two weeks ago a young kid at Mack, one week on the job, no union seniority or nothing, had a hand cut off in the press. Well, of course the big hassle was because they wouldn't allow the steward in the room with the foreman and the kid. The steward was helpless, because he didn't have representation in there—the kid wasn't a union member—you're not a union member for thirty days. And the guys walked out. Walked off the shift. They went back the next night, but all the international kept saying was, 'We'll take care of it, we'll take care of it.' The international vice president came down and everything. And even the guys I was talking with last night were saying, 'It's a terrible thing when you figure a steward getting physically abused by the labor relations guy'—and of course that's what their main concern was—and I asked what's going to happen to the kid? 'Ah, don't worry, workmen's comp says that they've got to find him a job. They'll find him a job somewhere.' This is an eighteen-year-old kid, without a hand now."

The worker who survives company intimidation is the worker tough enough and quick-witted enough to beat the foreman on his own terms. A black worker at Ford related the following incident to a black interviewer:

"Sometime in '67 or '68, I broke a finger. My foreman at the time gave me a gun weighing six pounds to work on the assembly line. I knew that I couldn't refuse to do the job, because I knew that if I refused, I could be fired. So I picked up the gun and jammed on the screw and dropped it. I made my feeble efforts. I didn't refuse the job, I made a feeble effort and dropped the gun and told him I couldn't do it.

"So he told me, 'You're gonna either do this job or get out of here.'

"I said, 'Peckerwood, if you don't get off from over my shoulder, I'll take that gun and cave your head in.' And there was one colored worker on the back of the car. He was the only one that heard me say this. So the foreman ran across to the superintendent. While he was going to get the superintendent, this, uh, *colored* man that's got in nineteen years at that time, says, 'You shouldn't have told him that you would kill him.'

"I said, 'Who heard it other than you?'

"He said, 'Nobody.'

" 'You gonna tell him?'

" 'No.'

" 'Well, what's the problem, then?'

"So by this time the superintendent came back over and he say, 'Goddamn it, I personally gonna see that you get your ass out on Michigan Avenue if you don't do what you're told.' And he stands six-four. So I stood up on the foreman's podium, which made me feel equal to his height, and I say, 'You big funky-ass peckerwood motherfucker, I'm not gonna kiss your ass or no other motherfucker's ass at Ford Motor Company to stay here.' I say, 'I lived before I came here, I'm gonna live after I leave here,' and I say, 'As a matter of fact, I'm leaving right now.' So I left— I went to Ford Hospital and got a restriction for my hand for six months. Work at my convenience. By a specialist.

"Now I told you this incident mainly to run down this thing about the superintendent. When I run off at the mouth at the superintendent like that, people on all four lines stopped and was looking and observing me and then this individual that heard me say I'd kill the foreman told me, 'You did right.' He say, 'I would have done the same thing but I got in nineteen years, and you only have two.'

"Well, I say, 'If the white man told you to get on your knees and suck his thing, what would you do?'

" 'Oh, oh!'

"I say, 'You just told me what you would do. All the man would have to do is be secluded some place, just you and he, and that's what you would do to preserve your nineteen years.' I say, 'You've got to have some principles that you're willing to die for. You've got to have manhood, something you're willing to die for.

"And he say, 'I-I-I'd kill him.'

"I say, 'No, you've already told me what you'd do. Now one thing you didn't take into consideration,' I say. 'Now, I think just as much of my two years here as you do of your nineteen. Why should you assume that my two years are nothing to me?' "

"But does the situation have to be that way," asked the interviewer, "where one individual worker takes off and fights back at the establishment?"

"No," said the Ford worker, "it doesn't have to be that way, but I'm afraid it's going to be that way for a long time."

Said the woman from Vernor Trim: 'Nobody knows if it isn't too late already—if there isn't too much dehumanization that has taken place and that we'll go the way of the Roman Empire or something, without any real force making any real change. And that's possible too, but there's no way of knowing that. All you can do is keep on fighting it.

"There is among workers a feeling that no change will really come about. That's what we're taught, from grade school on up. The you-can't-buck-city-hall attitude. It doesn't mean just city hall, it means everything. You just can't get out from under, and so in various ways you acquiesce to it. So what my attempt has been for a long time is just in very minute ways, not big grandiose declarations of any kind, or big political theories of any kind—not that I don't think they're important, but I don't think they're important to somebody who doesn't give much of a damn about anything—to therefore just discuss what goes on daily in the plant, how people have tried to fight. What successes, what failures have taken place. What it means. How to do things in the future. And a lot of it is just building morale. Just to let people know they really are human and they really do have a right to complain about those things. And how to do it so that you don't become completely dehumanized. And until that's done—I remember talking to one of the black guys in the shop who was telling me it sounds like what his attitude was for a long time toward the black community—that if the blacks feel they are human they can do something. And I think there might be some relationship with workers too that if they feel they're human, then they'll stand up for their rights more."

Up at the "Glass House," as the Ford Motor Company's world headquarters building in Detroit is commonly called, you hear a quite different story. The world headquarters building is air-conditioned, the floors vinyl-covered, the walls a checkerboard of glass and wood paneling. From behind his steel and gray vinyl desk, sitting in his blue vinyl and steel swivel chair, company spokesman Jerry Sloane told me: "The company is very proud of its safety record." He went on to say that the auto industry is, as a matter of fact, extremely safe in comparison to manufacturing in general, and that Ford has the best record in the auto industry. He gave me a copy of the most recent National Safety Council statistics, pointing out that while all industry lost an average of 672 days per million man hours worked, the auto companies lost only 179 days and Ford only 125 days per million man hours.

Sloane arranged an interview with Ford's medical director, Dr. Duane Block, who was able to shed further light on the Ford point of view. He explained that Ford had a four-point program that included: (1) selective job placement in keeping with health; (2) treatment and rehabilitation; (3) health maintenance; and (4) medical leave and medical control—which means giving each worker a physical every three years, and after he is forty every two years, and after he is fifty annually. "We're concerned about keeping Ford Motor Company employees healthy. We would like to instill in them a desire to maintain occupational health."

The real problems, he said, are in the smaller shops. "Most large industries like Ford—we think have a pretty good handle on this."

Under the Occupational Safety and Health Act of 1970, each state was given the opportunity to present a program, and if it was accepted, the state could assume all enforcement activities for its plants. So if state plans were approved, the law would be only as strong as each state wished to make it.

In Michigan, Detroit had its own Bureau of Industrial Hygiene, with jurisdiction over all workplaces in the Detroit area, a total of thirty-two thousand workplaces. If the attitudes of A. J. Kaimala, Detroit's senior associate industrial hygienist, were at all typical, enforcement would be weak indeed. Influenced, no doubt, by years of powerlessness, Kaimala made statements that bordered on the absurd.

"We've been controlling hazards for years," he said. "We don't look at these as *violations*, we look at them as *correction orders*." The companies, he said, "are cooperative. There is no such thing as uncooperative—some are less cooperative. They have budgets like everyone else. Everything is a matter of time. Extensive controls require much more time. I can't say noncooperative management really exists—sometimes you have to go there two or three times to get these things in. It's not that they don't want to do it, it's just that they forget. I don't think anybody deliberately does anything wrong.

"There's no fine on these correction orders," he explained. "They have to do it—they'll eventually do it."

"A lot of these things aren't life-and-death deals," he said. And he added, "This thing's been going on a hundred years. Who are we to shut them down in one hour?"

Kaimala was the traditional occupational health official, convinced that the kindly employers of Detroit needed only a gentle guiding hand to help them down the path of righteousness. He saw them as being like a balky horse, responding better to kindness than to the whip. "Between you, me, and the fence post," he said, "I don't think I'll live long enough to see this law [OSHA] enforced. You've got to sell your program. If you go in and say, 'Now, look, here's section 301, you're in violation,' the company will hire a lawyer and fight it for twenty, twenty-five years, because you've got to prove illness."

Nor did he believe in making inspections public records. "It just isn't done. You could just crucify a company. This is the business of the company. It doesn't become every John and Mary Doe's concern. It isn't a matter of suppression—it's a matter of not crucifying somebody."

Quite understandably, insurgent groups working for better conditions in the plants have seldom reported complaints to the bureau. Instead, they have attempted to work through the National Labor Relations Board. By law the NLRB is charged with protecting the worker's right to refuse to work under "abnormally dangerous conditions." Ron Glotta and Mike Adleman have represented worker's groups before the NLRB on several occasions after wildcat walk-outs resulting from deaths or injuries to workers. So far the board has rejected such cases. Why? Jerome Brooks, regional director of the NLRB, explained to me in 1970 that the board had refused to hear such an argument after Gary Thompson's death and the subsequent walk-out at Eldon Avenue—because the conditions at the plant were not *abnormally* dangerous. If the board stepped in every time there was an unsafe condition in a plant, he reasoned, "we'd be reaching into practically every shop in the country, wouldn't we?"

SCIENCE

The Drab and Slut of Industrialism

On a sunny summer day in 1930 the New Kanawha Power Company, a Union Carbide subsidiary, began excavation of a three-mile tunnel in Gauley Bridge, West Virginia. When finished the tunnel would carry water from the New River through a mountain to a hydroelectric power plant at Gauley Junction. Drawn by promises of better wages and steady work, Fayette County men left the coal mines to labor in the tunnel; from southern states men by the hundreds, most of them black, came north to work on the huge project, which at its peak employed a thousand men. No one dreamed that the dusty air in the tunnel would bring early death to perhaps half the workers.

It was six years before the tragedy was brought to public attention. In 1936 Miss Philippa Allen, a New York social worker who investigated the conditions at Gauley Bridge, testified before a

Congressional committee that contractors Rinehart & Dennis knew the rock they were drilling was ninety-eight to ninety-nine percent pure silica, yet they made no attempt to suppress the dust or provide protection for the workers. As a result, she said, "almost as soon as work was begun in the tunnel the colored men began to die like flies, because the percentage of silica in the dust they inhaled was so large. The ambulance was going day and night to the Coal Valley Hospital. As soon as a man died they would bury him . . . One colored boy died at four o'clock in the afternoon and he was buried at five o'clock the same afternoon without being washed. Why? Because the company did not wish an autopsy made, which autopsy would have uncovered the cause of his death."

George Robison, a tunnel worker, testified that most of the drilling in the tunnel was done dry (wet drilling is an essential means of controlling dust) because "otherwise they couldn't drill fast enough . . . The boss was always telling us to hurry, hurry, hurry." He said the migrants were forced to live in company houses, until they were too sick to work. Then the sheriff would evict them. "Many of the men died in the tunnel camps," Robison said. "They died in hospitals, under rocks, and every place else. A man named Finch, who was known to me, died under a rock from silicosis."

At first, workers didn't know what was afflicting them. Company doctors, said Miss Allen, told many they had "pneumonia" and others that it was "tunnelitis."

"It was a Mrs. Jones who first discovered what was killing these workers," she said. "Mrs. Jones had three sons—Shirley, aged seventeen; Owen, aged twenty-one; and Cecil, aged twenty-three—who worked in the tunnel with their father. Before they went to work in the tunnel, Mr. Jones and Cecil and Owen worked in a coal mine; but it was not steady work, because the mines were not going much of the time. Then one of the foremen of the New Kanawha Power Company learned that the Joneses made home brew, and he formed a habit of dropping in evenings to drink it. It was he who persuaded the boys and their father to give up their jobs in the coal mine and take on this other work, which paid them better. Shirley, the youngest son and his mother's favorite, went into the tunnel too.

"Mrs. Jones began to be suspicious when she saw the amount of sediment that was left on the bottom of the tub after she had washed the clothes of her menfolk. She asked the foreman about the dust and he said it was just ordinary dust and would not hurt anybody. Then one day Shirley came home and complained, 'Ma, I'm awful short-winded.' She said to him, 'Well, if you never feel no better, you'll not work no more.' He died nine months later, after telling his mother, 'Mother, after I'm dead, have them open me up and see if I didn't die from the job. If I did, take the compensation money and buy yourself a little home.' Within thirteen months Mrs. Jones's other two sons also died. Mrs. Jones filed suit against the Rinehart & Dennis Company for the deaths—hers was the first of two hundred suits asking a total of four million dollars in damages from the contractors. The same year *Business Week* noted that similar suits for silicosis claiming another thirty million dollars in damages had been brought against New York foundries. With so many claims being filed, the magazine said, 'employers don't know what to do.' Publicly, of course, they attempted to sound calm. Union Carbide declared it was 'very proud of its safety record everywhere.' Rinehart & Dennis claimed, 'We know of no case of silicosis contracted on this job.'"

Estimates of deaths among the four thousand workers who labored in the tunnel from 1930 to 1936 ranged from a few hundred to two thousand. Whatever the figure, the Gauley tunnel tragedy became a national scandal. Senator Rush Drew Holt of West Virginia labeled it "American industry's Black Hole of Calcutta." With public interest aroused, popular and scientific magazines began to write about conditions in the dusty trades throughout the country. Industry was indeed in a panic.

Scarcely a week after the Gauley Bridge hearings were adjourned, a group of industrialists met secretly at the Mellon

Institute in Pittsburgh to discuss the problem. The Mellon Institute was founded by Andrew and Richard Mellon in 1913 to "benefit American manufacturers through the practical cooperation of science and industry." According to Harvey O'Connor's book *Mellon's Millions,* published in 1933, Mellon fellows were supported by grants from individual companies to do confidential research by their sponsors. One institute project, for example, supported by a mattress manufacturer, "proved that the company's mattresses were 'most restful to the human form.'" In another, fellows supported by the Ward Baking Company discovered a secret process for bread-making which reduced by one-half the needed amount of yeast and sugar, resulting in increased profits for the company of a million dollars a year. "While the Ward stockholders rejoiced," wrote O'Connor, "nutrition experts shook their heads. White bread, they declared, depends on yeast for its vitamin content. If baker's bread is tasteless and lacking in nutrition, it is due partly to science's contribution in making available the use of cheaper grades of flour and savings in other constituents in the staff of life. They were reminded of Vernon Parrington's remark that science has become 'the drab and slut of industrialism.'"

In their hour of crisis after the Gauley Bridge disaster, the industrialists turned naturally to Mellon. Out of their secret meetings at the institute came the formation of a new organization, the Air Hygiene Foundation. The purpose of the foundation, according to the confidential report of the proceedings, would be "to better dust conditions in the industries." But, in fact, the men of science and industry had acted more from concern for their own survival than for the survival of working men. "Because of recent misleading publicity about silicosis and the appointment of a Congressional committee to hold public hearings," noted the report of proceedings, "the attention of much of the entire country has been focused on silicosis. It is more than probable that this publicity will result in a flood of claims, whether justified or unjustified, and will tend toward improperly considered proposals for legislation." To forestall such calamity, the foundation planned a public relations campaign to "give everyone concerned an undistorted picture of the subject."

Among the members and trustees of the fledgling foundation were several highly reputed scientists and important public officials, including A. J. Lanza, M.D., formerly chief surgeon of the U. S. Public Health Service, then medical director of the Hydraulic Steel Company; Philip Drinker, a Harvard professor and perhaps the most widely respected industrial hygienist in the country; R. R. Sayers, M.D., of the U. S. Public Health Service; and Daniel Harrington of the U. S. Bureau of Mines. To companies that had been working with the Mellon Institute, apparently nothing appeared to be unusual in such an affiliation.

In its early years, as the foundation worked actively to promote its goals, its spokesmen were widely quoted in popular trade publications. In a 1936 issue of *Coal Age,* for example, Alfred C. Hirth of the foundation was quoted as saying that "silicotics are rare compared with men driven from their jobs by shyster lawyers . . . Speaking generally, with a little persuading, able-bodied men with good jobs are convinced that they are afflicted with an insidious lurking disease and that their present employment means a premature death." Another foundation member, attorney Theodore C. Waters, blamed *doctors* for causing inflated silicosis claims. "In many instances employees have been advised by physicians, untrained and inexperienced in the diagnosis and effect of silicosis, that they have the disease and thereby have sustained disability. Acting on this advice, the employee, now concerned about his condition, leaves his employment, even though that trade may be the only one in which he is able to earn a living." The "resulting hardships" could be avoided, concluded Waters, if the man had been "properly advised that he had suffered no actual disability and that the risk of continuing in his employment was negligible."

Whether industry was impressed by the logic of such arguments, or wished to avail themselves of the foundation's various other services, the foundation's membership grew steadily. In its second year, the first year it released figures, member companies totaled 192. By 1940 there were 225 member companies, including many well known and giant companies such as American Smelting and Refining, Johns-Manville, United States Steel, Union Carbide, and PPG Industries. A more recent list of 400 members included Gulf Oil, Ford Motor Company, General Motors, Standard Oil of New Jersey, Kawecki Berylco Industries, Brush Beryllium, Consolidation Coal, Boeing, General Electric, General Mills, Goodyear, Western Electric, Owens-Corning Fiberglas, Mobil Oil, and Dow Chemical.

At the foundation's fifth annual meeting, C. E. Ralston, chairman of the membership committee and safety director for Pittsburgh Plate Glass Company, made a plea for new members, reminding those present of the virtues of foundation membership: "A survey report from an outside, independent agency carries more weight in court or before a compensation commission than does a report prepared by your own people. One of the brilliant features of AHF is this: it is a *voluntary* undertaking by industry to protect industrial health. And where industry attacks a great social-economic problem voluntarily, there is no necessity for government to step in and regulate."

Over the years the foundation, which changed its name to the Industrial Health Foundation, continued to provide a hospitable forum at its annual meetings for the somewhat special opinions of its members. At one meeting, for example, Dr. E. A. Irwin, then medical director for Ford Motor Company, told foundation members that almost all heart attack victims can and should return to work. And getting them back to work would be easier, he thought, if it weren't for too-liberal compensation allowances in some states that "make for the acceptance too easily of the idea that working conditions are the cause of heart attacks."

Dr. Irwin's views may have been mild, as foundation affairs go. At an earlier meeting, Dr. William Shepard, third vice-president of

Metropolitan Life, told his listeners that too much emphasis was put on "the stress of overwork as a cause of heart attacks, and of such ailments as arthritis, high blood pressure, ulcer, gout, asthma, bronchitis, nervousness, and various neuroses." More blame should be put on "the inexorable aging process which begins at birth," he declared, "or on improper habits. And bad habits," he said, "are probably more common among loafers whose major object in life is to escape stress of any kind."

Former foundation president Dr. R. T. P. deTreville also expressed his concern about the problem of aging among workers. As he explained to an undoubtedly appreciative audience at the 1969 annual meeting, "IHF programs for many years have stressed that there is no reason to believe that prevention of specific occupational diseases will necessarily advance employee health, that is, by preventing chronic diseases common to the general population . . . There will be many environmental hazards found which require monitoring and control," he admitted, "but by far the largest source of morbidity, mortality, and noneffectiveness in the work force will be found to be the nonoccupational diseases which are prevalent in the aging general population."

Foundation members have boldly gone where others feared to tread, asking questions that no one else could even have thought of. Carl G. Staelin of Owens-Corning: "Now let me ask a question about this widespread interest in pollution. Is this good? Is it a good thing that there is now a widespread interest in and demand for pollution abatement?" He went on to say that from the "social point of view" admittedly it is good. He pointed out that although for industry such concern often creates "serious problems," this time "people are deadly serious about ending pollution." Industry might as well realize, he sighed, that "the time for paying lip service to pollution control is past and . . . the time for action is now." He suggested that other companies may want to follow the course of action that Owens-Corning had found so profitable. In a statement more than ordinarily candid about the benefits of IHF membership, Staelin said: "We occasionally hear complaints or accusations that under certain conditions or in certain situations glass fibers present a health hazard. We learned long ago that the only practical defense against such accusations is to offer proof that our fiber glass products do *not* present a health hazard. In this respect, we have made considerable use of the Industrial Health Foundation." Numerous industries, notably coal, lead, petroleum, asbestos, and textiles, have sought the same sort of succor from the IHF.

To hear one of the foundation's former executives talk, it is no less than God's work that they have undertaken. "Our concern is for the *people* and that's what we're in business for," IHF vice-president Harry Bowman told me in 1970. "It sounds almost missionary, doesn't it?" Hostile and suspicious at first, Bowman grew cordial. He enthusiastically described the foundation's staff of twenty-five as a "small but dedicated group of people," doing work that is "truly ecumenical," he said, patting my knee reassuringly.

Although the foundation's primary purpose is "research and education," Bowman said, the foundation has on occasion testified "on a technical basis" on worker health legislation in such places as Boston, New York, Philadelphia, Kansas City, "and of course, Washington." On one such occasion, in 1968, Dr. Paul Gross, IHF director of research, testified before the Senate labor committee on the proposed coal mine health and safety act. Gross, a lung pathologist, had never examined a coal miner himself. Still he testified that he believed that of fifty thousand disabled miners, only thirteen hundred to five thousand were disabled from occupational lung diseases and that the other forty-five thousand were suffering from other causes, chief of which he believed to be cigarette smoking.

Senator Harrison Williams, committee chairman, found the testimony hard to believe: "Are you saying that in the forty-five thousand residual figure—assuming fifty thousand is accurate—the disability is not related to coal mining at all?"

Gross replied (on a technical basis, of course): "Only insofar as the inhalation of dust from an epidemiologic point of view causes a slight increased prevalence of chronic bronchitis and emphysema among miners than it does among non-miners." (Since the law was passed in 1969, more than 230,000 miners have been compensated for total disability due to black lung.)

The foundation has also testified in numerous workmen's compensation cases, when called upon by any of its six hundred companies—and, as Bowman said "when it is a question of *objective* scientific medical aspects. This is the kind of *professional* testimony we can provide—expert testimony in the fields of our competence. We've had people contend that they've contracted asbestosis, silicosis, or fibrosis from a given work environment when you know what is in the environment because you've measured it."

Consider, for example, the foundation's role in a 1961 compensation case. When two employees of the Hall China Company, an Ohio pottery manufacturer, filed claims against the company charging disability caused by silicosis, the company called on the IHF for assistance. The foundation responded with a report quoting an outdated and discredited U.S. Public Health Service study which said that if dust levels in potteries are kept below four million particles per cubic foot "new cases would not develop." Then the IHF report presented results of two dust surveys made eleven years apart at the Hall China plant, concluding that because the dust samples were below the four million particles per cubic foot level, "the dust counts at your plant have not been sufficient to constitute injurious exposure since 1945." The foundation's findings could hardly be characterized as objective or even professional testimony. As Dr. Thomas Mancuso, then chief of the Ohio Division of Industrial Hygiene, commented, the foundation's argument assumed that dust concentrations at the plant remained, day in and day out, at the same level measured in a limited number of samples taken in surveys eleven years apart. The foundation's conclusions were particularly difficult to accept, he concluded, since fifty-six Hall China employees had already been compensated for silicosis, and seven more had claims still pending.

The foundation's activities have been viewed with some skepticism in recent years by many occupational health professionals, but most are reluctant to say anything against them publicly. "Nobody's ever denounced the foundation or exposed it in any way," said Dr. Mancuso, who has had his eye on the outfit since his encounter with it in various Ohio compensation cases. "There's no tangible or clearcut evidence. We all have our personal reservations—things that we have picked up."

"It's how the game is played," Dr. Mary Amdur told me. "You're still respected even if you make outrageous statements. Once someone is established, he's not condemned outright, particularly if he is paid by industry." And another researcher, though he said of the IHF staffers that "most people who are good scientists don't even consider them seriously," also said: "You don't see them attacked—it's the tradition of scientific literature. You don't get emotional; you answer with another paper." Dr. Harriet Hardy, characteristically candid yet unwilling to condemn the IHF people out of hand, said: "Good things can come from the devil."

The foundation's most powerful ammunition in recent years has come from its research division, headed by Dr. Gross. Gross seemed the stereotypical German scientist, with a heavy accent and a short, white, pointed beard and wearing the usual white lab coat. He was gracious, charming. The laboratory and Dr. Gross have been an integral part of the foundation since 1948. In those early years, the laboratory was part of Mellon Institute, and Dr. Gross, as a senior Mellon fellow, conducted research for industry on the same basis as did other Mellon fellows. In 1968, after the institute became part of Carnegie-Mellon University, the building in which the lab was housed was leased to the University of Pittsburgh, apparently with the understanding that the lab came with the building. Exactly why they moved is unclear, but the laboratory and Dr. Gross became a part of the University of Pittsburgh with the blessing of the school administration. Dr. David Minard, chairman of the School of Public Health, said later that the affiliation had met with his approval. Though the laboratory continued to be wholly funded by the foundation, he claimed this did not compromise the university. "It is no different than the kind of research that would be done from government funds. IHF gives a grant to the university which is handled like any other grant. I for one feel that it's not a matter of being in bed with industry any more than being in bed with the AEC or HEW or the military." (The school's research funding came primarily from those three sources.)

Minard, apparently sensitive about criticism of the association, alternately praised the foundation and minimized its importance. He said on the one hand that the affiliation would be coming to an end and that the lab's resources "never amounted to more than fifteen percent of our total research effort." He said on the other hand that the foundation was a perfectly legitimate, nonprofit, tax-exempt foundation, its purpose being "exactly the opposite of trying to cover up the problem. It's trying to get the solution. To call the IHF an adversary to labor is errant nonsense—it's really the laboring man, the worker, that the foundation is set up to help." In an aside which sounded strangely familiar and also inadvertently revealing, he added that the main problems that industry faces now are "the chronic degenerative processes of aging and to what extent occupational exposure may accelerate these diseases."

Others in Minard's department had disagreed and had fought the foundation's affiliation with the university. One member of the department said he opposed it because "I think they can make mileage out of the university. Apparently it [the decision] came from very high up. About four of us in this department—about one-third—opposed it. When we opposed it, we bucked right against the administration. I just don't really like it. Their goal is to protect member firms. Their interests are not those of the worker."

Because of scientists' traditional reluctance to air anything but "scientific" disagreements publicly, and because the scientific community lends acceptability to anyone who is a member, those with the proper credentials can abuse and twist the meaning of "science" to suit their own purposes with some impunity. Such abuses are probably seldom consciously contrived, schemed, and plotted, but instead occur as the result of slow erosion of conscience, until to the scientist involved the most dastardly of actions seems to be fair and honorable. And if he feels trickles of doubt from time to time, he is quickly reassured by those around him. The principals are constantly engaged in patting one another on the back for their virtues, a practice engaged in—it seems—in inverse proportion to the deserving character of the practitioners. So many and so outrageous are the lies that each must make within himself that in time nothing appears outrageous. At IHF meetings, a Johns-Manville executive can boast happily of how his company solved its public relations problem, while at the same time Johns-Manville workers are dying of asbestos-caused cancer in epidemic proportions. An attorney can actually propose to the members that they plan for industrial disasters, deciding beforehand which people may be considered expendable (the workers) and which to be protected (executives, children, and Red Cross volunteers).

Science lends an air of respectability to the Industrial Health Foundation and the foundation gives credence to the views of its members. The oratory at the annual meetings is often too absurd to be really dangerous, or to be appreciated by any but the true believers. The dangerous work is done in the foundation's laboratories under the direction of medical doctors.

Since 1948, when he became director of the foundation's laboratory, Dr. Gross has studied and written prolifically about dusts of all sorts—silica, diatomaceous earth, glass dust, aerosols, fluorspar dust and radiation, enzymes, fiber glass, aluminum, and asbestos. By 1972 he had published more than two hundred papers, a fact that gave him some pleasure. In recent years, with money coming partly from the asbestos industry and partly from the U. S. Public Health Service, he has worked primarily on research relating to asbestos dust, or, as he defined it, work "to determine the dust content of human lungs—the nature of the dust, and with particular emphasis on the mineral fiber."

Since the early 1900s, asbestos had been associated with an emphysema-like disease called asbestosis, although the disease was not widely recognized or reported until around 1930. Scattered reports of lung cancer associated with asbestosis appeared in the literature beginning in 1935. Twenty years later the relationship between asbestos and lung cancer was clearly established when a British scientist found an incidence of lung cancer among British asbestos workers eleven times as great as that in the general population. And then, in 1960, asbestos was linked with still another disease, mesothelioma, a heretofore rare cancer which attacks the lining of the lung and body cavities. It is always fatal.

That same year a group of mesothelioma cases were reported in South Africa, in the vicinity of the Cape of Good Hope asbestos mines. Most of the victims were not occupationally exposed to asbestos but had simply lived in the vicinity of the mines, developing the cancer twenty to thirty years after initial exposure. Soon after, a similar epidemic was reported in Pennsylvania in the vicinity of an asbestos manufacturing plant. Because of these "community cases," as they are called, no one could be sure just how little exposure was necessary to produce the disease. But in 1963 a scientist named J. G. Thomson published the results of a study in which he claimed to have found "asbestos bodies"— microscopic particles of asbestos—in the lungs of twenty-five percent of the inhabitants of Capetown, South Africa. Thomson's findings of widespread exposure to asbestos dust among the Cape population were viewed as potentially a cause for some alarm.

Similar reports followed. Asbestos bodies were observed in autopsied lungs in one-fourth the population of Miami, forty-one percent of the population of Pittsburgh, and forty-two percent in San Francisco. A review of the literature on the hazards of asbestos prepared by Litton Systems for the National Air Pollution Control Administration concluded (though the conclusion was suppressed in the official version) that "asbestos is an air pollutant which carries with it the potential for a national or world-wide epidemic of lung cancer or mesothelioma of the pleura or peritoneum." The report estimated the number of people exposed to asbestos to be: 110,000 asbestos workers; three and a half million people working in areas where they may be exposed to small quantities of asbestos dust; and fifty to a hundred million people in the general population who have breathed or will breathe enough fibers to show "asbestos bodies" at autopsy. Asbestos researcher Dr. Irving Selikoff commented that while the presence of the fibers in some lungs may not necessarily mean cancer will follow, the potential is there, and that "is what is worrying all of us." The implications, both medically and for the asbestos industry, were grave.

Dr. Gross's laboratory took issue with Thomson. "Not with his findings," Gross said, "but in regard to his interpretation of his findings. We believed that these bodies, which [Thomson] called asbestos bodies, were not necessarily asbestos bodies—that other fibers were capable of producing the same substance as were asbestos fibers. And we subsequently proved this with animal experiments in which we introduced fibers of various types into the lungs of hamsters and found in these hamsters, as well as in guinea pigs, the same structures that, in response to the presence of non-asbestos fibers, Thomson had called asbestos bodies. We gave the name of 'ferruginous bodies' to these structures because the fibers were encrusted by an iron-containing protein." Among a series of papers Gross wrote in pursuit of proving his theory—and, incidentally, in defense of the asbestos industry—was one titled "Pulmonary Ferruginous Bodies in City Dwellers."

The paper, a two-page report published in the August 1969 issue of the *Archives of Environmental Health,* was to be a fundamental document, upon which Gross would build the blocks of his theory that the "ferruginous bodies" found in the general population came from sources other than asbestos. The report was summarized as follows:

> Chrysotile [a type of asbestos] which comprises more than 90 percent of the asbestos used in this country, has a characteristic electron diffraction pattern because of its unique, hollow, tubular, crystalline structure, as seen under the electron microscope.
>
> On the basis of the electron diffraction pattern chrysotile was decisively excluded as a constituent of the cores of all 28 ferruginous bodies isolated from lungs of urban dwellers not occupationally exposed to asbestos.
>
> This exclusion is considered highly significant because if the ferruginous bodies in the above city dwellers had been caused by the inhalation of asbestos dusts, then some of the cores should logically be composed of chrysotile.

Certainly if Gross's work was correct, the findings of Thomson and others might be put into some doubt, or at least the issue would be thrown into confusion. But according to two other scientists, working for the foundation at the time, one of them co-author of the paper, the conclusions were not justified. Martin Haller, a co-author of the study and a Mellon fellow, who did the electron microscopy work, disagreed with the results and conclusions of the study. Studies made at Mt. Sinai Hospital in New York City, Haller said, done since the Gross study, proved without a doubt that the fibers of city dwellers *were* in fact asbestos. In the Mt. Sinai studies, he said, "dozens, hundreds" of fibers were taken for identification from each lung, while in Gross's study, "we ended up usually with a single fiber from one lung. I believe they were premature in reaching conclusions I did not think were justified by the facts. The writing was done by Dr. deTreville [who also appears as a co-author] and edited and published without my seeing the final form. I will admit doctors Gross and deTreville had a tendency to do this, run away with everything. I didn't realize until afterward that they would write in an exaggerated manner and not based on what we had done."

Haller said he didn't want to be pictured as constantly at odds with the foundation. He said he didn't discuss his disagreement with either Gross or deTreville at the time because he thought "it wasn't that important," but he said he had "private reservations."

"I didn't think it was necessarily the best way to go about it," he said.

Dr. Michael Utidjian, a fellow with the foundation at the time, was also involved in the "ferruginous body" studies, and had been the main author of an earlier study, with Gross and deTreville, in which he found "ferruginous bodies" in ninety-seven percent of one hundred autopsied lungs of Pittsburgh citizens. It was Utidjian's samples that Haller was given to identify. "While I was identifying these ferruginous bodies in sample after sample," he said, "Gross would come in and we would discuss them. He said, 'So far as I know, only chrysotile will look like that'—chrysotile fibers have a distinctive appearance; they look like a jazz drummer's wire brush, frayed and split out at the ends. Gross told me these at least are most certainly chrysotile." Haller was able to use about twenty of the hundred specimens. "It's a fairly hit-and-miss procedure," says Utidjian, "and of those twenty specimens he wasn't able to identify them as anything else [but asbestos]. DeTreville and Gross published a paper saying *none* of those were asbestos-based; in other words, 'Get asbestos off the hook.' People in the know saw through this false reason." It was because of this incident that Utidjian left the foundation, though he had been "pretty disillusioned" for some time.

"I came from Britain," he explained, "where medicine is completely socialized. The IHF was presented to me as completely impartial. I found this was in fact not the case. The choice of what part of the work is published was very much dictated by deTreville. If it was financially inimical to industry, it didn't see the light of day." His first project for the IHF was a study on the threshold limit value (TLV) for antimony. "The way deTreville put it to me, he said the ACGIH [the American Conference of Governmental Industrial Hygienists] were wondering whether the TLV for antimony had been set too low." Utidjian spent three months researching the literature and concluded that the existing TLV, which was based on the TLV for arsenic, should not be raised (or relaxed), and in fact that there was a case for *lowering* the standard for some compounds. "When I told deTreville," he said, "deTreville replied: 'You mean to say you can't find any *smidgeon* of evidence that the TLV could be raised?' And I said, 'Nope.' Having heard this, the whole subject was dropped."

DeTreville appeared regularly at ACGIH meetings to produce evidence on standards, said Utidjian. "He was always on the side of industry if there was any doubt. He just has blinkers on. Industry has always been able to find its apologists and expert witnesses."

The foundation's influence with the ACGIH was considerable— and important to industry. Until the enactment of the Occupational Safety and Health Act of 1970, the ACGIH was the quasi-official standard-setting body for toxic gases, dusts, fumes, and vapors. Its standards were used by many state programs and the federal Walsh-Healey Public Contracts Act as well, and the ACGIH standards overwhelmingly became the standards in the 1970 federal act. Although supposedly an unbiased scientific committee, in fact it was organized and at this writing was still ruled by one man, Dr. Herbert Stockinger. A cigar-smoking, self-confident man who appeared to be in his sixties, Stockinger was chief of toxicology at the National Institute of Occupational Safety and Health (NIOSH), a federal agency located in Cincinnati; a member of the IHF's Chemical Toxicological Committee; and "a great friend of" deTreville's. Singlehandedly, Stockinger selected members for the ACGIH committee. Among them was Dr. Gross, since the mid-sixties head of the important committee on dusts. Stockinger saw nothing wrong with the way he ran his committee, or the fact *that* he ran it. He ought to run it, he figured, because he's been in the field forty years. "I kind of know my way around," he told me. In that time he claimed he'd seen "no evidence of bias" from industry. And anyway, he added, "I'm chairman of the committee; I can accept their opinions or not. The way you operate best in this area is to accept the opinions of those best informed."

Such as Dr. Gross. Stockinger saw no bias, of course, because Gross's bias was his own. But that it was there was abundantly clear. The ACGIH fibrous-glass dust standard, for example. The documentation for that standard—set at 10 and labeled an "inert" or harmless dust in the ACGIH's 1971 edition of "Documentation of the Threshold Limit Values"—might pass as a straightforward scientific argument, but only to a very uncritical or very sympathetic eye. The bulk of the "documentation" rests on Gross's *unpublished data,* "personal communications" from Owens-Corning Fiberglas Corporation, and "unpublished data communicated to Repository of Anonymous Industrial Hygiene Data, Ind. Hyg. Fndn. of America (1968)"! Most recently, researchers who believe in publishing their data have produced mesotheliomas in experimental animals with fiber glass—but the ACGIH showed no sign of changing its standard.

And the committee may continue to be influential in the future, even though its activities have been officially bypassed under the new law. Stockinger declared that the ACGIH will continue to operate and "is going to be an opposing voice for what NIOSH has put out." He mentioned, for example, one substance for which NIOSH had "seen fit" to recommend a lower, stricter standard, even though Stockinger's committee had "seen no reason to change it." "This has got to be adjudicated, eventually by the OSHA committee," he said emphatically. "We will not take carte blanche the recommendations of NIOSH!"

In addition to his dismay with the foundation's role in standard-setting, Utidjian soon found even stronger disillusionment with his employer. He recalled a memorable incident soon after he had joined the foundation. A pipe manufacturer in the Pittsburgh area phoned deTreville with an emergency plea for help. The plant had just opened using a spray process with highly toxic epoxy resin materials. Within a few days of operation two men became seriously ill and were hospitalized. The company didn't know what to do. After deTreville had been told about the situation, said Utidjian, "he got wildly excited and said, 'Come on, let's go out there!'" Utidjian and deTreville drove to the plant and observed the process. "The company apparently had been doing the same process in Tennessee for several years without any trouble, but there they were doing the spraying out of doors. Here it was in what in England we call a quonset hut—a tightly enclosed area— you couldn't see the workers from across the hut, there was so much dust and fumes and stuff. The materials—epoxy resins—are potentially very dangerous to the lungs."

After a quick view of the plant, the two drove to the hospital to see the workers. "One of the men nearly snuffed it and just pulled through, and another was seriously ill with chemical pneumonitis," Utidjian said. "While we were driving back I said, 'Well, obviously the process has to be stopped immediately.' DeTreville was horrified. He said, 'You can't do that! You can't stop industry.' He said, 'Besides, if we stop it now, we can't study it!'

"From that time on," Utidjian said, "I started looking for another employer."

Nonplussed by Utidjian's disapproval, Gross and deTreville went right on publishing. In a paper entitled "Fibrous Dust Particles and Ferruginous Bodies" Gross drew on the conclusions of the earlier paper to exonerate chrysotile asbestos, the mainstay of American manufacturers. "Electron diffraction analysis of the central cores of 28 randomly selected ferruginous bodies from city dwellers indicated a crystalline structure in all of them. This fact eliminated glass and ceramic fibers as being responsible for the formation of these particular bodies. By exclusion . . . this left asbestos as the probable constituent . . . However, because chrysotile had been decisively excluded as being involved . . . and since more than 90 percent of the asbestos used in this country is chrysotile, an enigmatic situation prevails." Gross concluded by suggesting the fibers in the lungs in this new study—which included seven city dwellers, two long-time employees of a fiber glass factory, and one of an asbestos products factory—may have come from exposures to talc dust.

In another notable study, in which Gross exposed rats to chrysotile dust, thirty-one percent of them developed lung cancer. He concluded that the cancer should be blamed on trace metals from the hammer which was used to pound up the dust! To a layman the suggestion is absurd on its face. But the result, as another researcher said, was to create a "red herring, which is the way he works. That's his technique of defending industry. He always raises these flimsy questions on the basis of the scantest of evidence, which forces you to spend money and do more research on something that's well proven."

What is more outrageous is that much of the funds for Dr. Gross's red-herring research came from Public Health Service grants and from contracts mostly with the National Institute of Occupational Safety and Health (NIOSH), and its precursor, the Bureau of Occupational Safety and Health (BOSH), agencies that suffered from inadequate funding and could scarcely afford to waste money on such enterprises.

Since 1963 the IHF has received at least $1.3 million dollars from Public Health Service grants and contracts, constituting about a third of the foundation's support. The foundation also found a great deal of sympathy for their proposals at BOSH. Lewis Cralley, in fact, head of the division of epidemiology and special services at BOSH, was not only in agreement with the IHF philosophy, he allowed the foundation to publish a book he had co-authored, *Industrial Hygiene Highlights.*

The old-guard Public Health Service officers at BOSH "had this thing about trace metals," said Dr. William Johnson, chief of the medical investigations branch of the division of Field Studies and Clinical Investigations for NIOSH. "It's a stalling technique," added Dr. Joseph Wagoner, director of the division. "It may not have been intentional, but that was the effect." And so Paul Gross, whose imagination conceived of a number of variations on the trace element theme, found the bureau a willing source of funding.

In fact, NIOSH grant director Dr. Alan Stevens saw no reason not to continue funding the IHF if it proposes projects of sufficient merit. And "these particular research projects, as designed, and the competence of scientists such as Paul Gross, are strong enough to merit funding," he said.

"We would take into consideration whether there's any problem of conflict of interest," he added, but Stevens' idea of conflict of interest is a rather strange one. Although the review committee turned down an IHF request for subsidization of their *Industrial Hygiene Digest* publication, because "they could and should be able to do it themselves," Stevens still thought "it might be in NIOSH's interest to subsidize it." (The *Digest* is full of interesting scientific revelations such as the fact that "coal dust is not the cause for breathlessness in coal miners, when it occurs, and . . . attempts to relate the two are sociologically rather than scientifically based.")

All of the concern about trace elements in asbestos and what exactly about asbestos made it cause cancer ultimately proved to be superfluous, as it became apparent that all kinds of asbestos, including chrysotile, will cause lung cancer and mesothelioma. And regardless of which variety and which trace elements might be present, what is most important is that worker and public exposures to the dust be controlled. Cralley's division at BOSH could have discovered that and taken steps to alert the Labor Department years ago, said Doctors Wagoner and Johnson. Their division at NIOSH had begun an asbestos study in 1964, but when Wagoner and Johnson inherited the division in 1971 they found to their amazement that none of the data collected for the study—two thousand asbestos samples and one thousand X-rays—had been analyzed and nothing had been published. As the two young doctors began to analyze the data they were staggered by what they found.

"I took home fifty death certificates to analyze," Dr. Johnson said, "and all fifty were from asbestosis, occurring in the 1960s. Undoubtedly a lot of their exposure was in the 1940s and '50s. They were dying a very short period of time after they terminated employment. There was no doubt in our minds that industry knew there was a serious health problem. It didn't take any sophisticated analysis to know—it seemed like it was a very obvious problem."

"If they had looked at the data," said Wagoner, "the answers were there. It doesn't take a genius to figure out these things. The important thing is what are the questions that are being asked."

Whatever their motives and intent, Doctors Gross and deTreville and their counterparts in government clearly lost sight of the crucial question: What was happening in the plants? While they contented themselves with measuring trace elements and counting fibers in the lungs, workers were dying of asbestos-induced disease in great numbers. Why Cralley never finished the epidemiological studies that he had begun, Wagoner didn't know, although he said: "I have my moral feelings about it. All I know is, it didn't come out. If it had come at a much earlier time, we would have saved a hell of a lot of lives."

Two hundred thousand workers are presently exposed to asbestos and there are another eight hundred thousand ex-asbestos workers, Dr. Wagoner said. Of those, based on studies he and Johnson have completed, "we can anticipate three thousand excess respiratory, cardiopulmonary deaths and cancers of the lung—three thousand excess deaths *annually* for the next twenty or thirty years. That turns out to be ninety thousand deaths."

While Dr. Gross was peering through microscopes at "ferruginous bodies" something grim was happening in asbestos plants across the country. There, no matter what Gross was able to "prove," workers were suffering in epidemic proportions from asbestosis and mesothelioma.

In 1972 Johns-Manville Corporation was the largest American producer of asbestos products. Its Manville, New Jersey, plant, employing twenty-one hundred workers, was the largest of the company's more than fifty U.S. operations. Bob Klinger, vice president of the United Papermakers and Paperworkers Union, Local 800, which represented J-M workers, said the union first began to suspect problems at the plant in the 1960s. "We weren't scientific people," he explained, "and we really didn't know what to suspect or what one would say is reasonable, except that we had a lot of people dying of cancers, a lot of people dying of coronaries, a lot of people taking disability retirement. All of these factors alarmed us."

Dr. Maxwell Borow, a local surgeon, began noticing an extraordinary number of cases of the rare and always fatal cancer, mesothelioma, starting with his first diagnosed case in 1963. By 1967 he had documented eighteen or nineteen cases, he said. He wanted to prepare an exhibit to show at medical conventions to alert other doctors to the disease, but, he said, "We couldn't get any funding for it." First he approached Johns-Manville, but the company refused to support the project. "They said they weren't prepared at that time to accept the relationship—that there was a relationship between asbestos and mesothelioma." Borow found other local industries also unwilling to provide financing, and finally he wrote to the union, which provided the money for the display.

In 1970, when the union went on strike, partially to protest hazardous working conditions, Dr. Borow's display was set up at the union hall. For many workers the exhibit was an eye-opener. "I guess most people really got aware of it then," said Dan Maciborski, a lathe operator at the Manville plant. "That's when everyone found out what it can do to you."

Two years later, Dan Maciborski found out that he had the cancer himself. I interviewed him in April 1972, just a few months after he learned of his diagnosis. Maciborski was reared in the Pennsylvania anthracite fields, and his father, a coal miner, had died of black lung disease. "I came out here," said Maciborski, "because my brother came out here with some friends. They got jobs, so I figured I'd leave the anthracite region myself, figured it would be safer, you know. But I guess it didn't turn out that way." Nor did it for his brother, William Maciborski, who suffered from asbestosis.

"In 1941, when I got a job there at J-M," said Dan Maciborski, "the front lawn was filled with men and they just picked out who-ever they wanted. Everyday they'd come out there, sit down, 'Okay, you, you, come in for interviews.' All they did was interview you and tell you it was *hard* work. And it really was. But they never told you that it was bad for you or hazardous, but they just said it was hard work. And it was back-breaking work. But I didn't mind working. Hard work didn't bother me—I was used to hard work, you know."

Maciborski, forty-nine years old when I saw him in 1972, was hoping the doctors would be able to arrest the cancer. He said he tried not think about it, "but it's always there." He died a year later.

The usual progress of the disease, Dr. Borow explained, is this: "Initially it starts out, like in the abdominal cavity. The only places you'll see tiny nodules—they look like tapioca pudding nodules—will be studding the peritoneum, both the abdominal wall and on different organs, the liver, the intestines. And these patients usually come to the attention of the doctor by virtue of a fluid—these little nodules secrete a fluid and these people come in with extended abdomens. Then as the disease progresses, these nodules get larger and then coalesce, and start forming solid masses. When you cut through it, this big mass—this tumor—there are the organs imbedded inside. They are not invaded, particularly, they're crushed. They're all crushed by virtue of this big mass, as if you had poured in cement."

Borow and his associates had identified more than sixty cases of mesothelioma from Johns-Manville by 1972. "Everyone is agreeing that this disease is just starting to show itself," he said. "Why, I don't know. But it's going to continue to show itself, for another twenty or thirty years, because it takes that long before the cancer shows itself."

Asbestosis, too, was a major problem at the plant. Union officials showed me state compensation figures for 1970—which they said was not a normal year because of the five-and-a-half-month strike. There were 170 awards against J-M, resulting in $370,530 in payments to afflicted workers. Figures for 1969 showed 285 cases. Under New Jersey compensation laws, workers were required to file every two years in order to continue receiving compensation. And there was a five-year statute of limitations as well. "If a man does not file for workmen's compensation for dust," explained Bob Klinger, "and he's away from exposure for a period of five years, he's not eligible. The company's scot-free. A lot of the old-timers, that we've contacted with our study, they're very bitter. Very bitter against the union, the company, everybody. Because here they are, they're diseased, they got half a set of lungs and they didn't get one cent for it. They're very bitter."

Dan Maciborski's older brother, William, was partially disabled by asbestosis in 1972 and drawing seventy-two dollars a week unemployment while he awaited the results of his most recent compensation case. He said, "Here they got my brother in a bad situation. And he's not the only one. My brother-in-law died from it.

There are more people sick with this than has been brought to the surface. A lot of them were dead and buried and never knew what they had, before they brought this out."

Union officials said that since the strike in 1970, the company had done a "good job" of cleaning up the plant. "But that's not going to eliminate the problem we have," said Klinger, "because people have already been exposed and have already been contaminated. No matter how much money they put in now, this generation of workers is contaminated."

Gross and deTreville finally lost out to the facts, with regard to asbestos, but they won the victory of delay. Bowman and deTreville have since left the foundation for other employment. Gross has retired, and Dr. Daniel Braun, former medical director of U. S. Steel, has been appointed IHF director. The work of the foundation continued, perhaps with slightly more sophistication under Braun, but with the same intent and effects as before.

One of the foundation's recent research efforts, directed by Braun, was on behalf of the American Textile Manufacturers Institute (ATMI) and concerned byssinosis. Byssinosis, or "brown-lung disease," has been recognized as an affliction of cotton mill workers since the early 1900s, but until recently few studies had been done in this country. A flurry of reports appeared in the literature in the early forties, but interest in the disease flagged after 1945, when a report published by the U. S. Department of Labor said byssinosis was not a problem in American cotton mills. In 1968, however, the specter of widespread disease in the mills was renewed when a research team led by Dr. Arend Bouhuys of Yale found a high incidence of the disease at cotton mills operated by inmates of the Federal Penitentiary in Atlanta. The study revealed that twenty-eight percent of workers in the carding and spinning rooms showed symptoms of the disease.

In the first stages of byssinosis, workers returning to work on Monday mornings feel tightness in the chest and shortness of breath, and they cough—a reaction to the cotton dust they breathe.

Later symptoms persist and can become permanent and disabling. The alarming thing about the report was that the disease was common despite the fact that the mills had "modern textile equipment and good general hygienic conditions." The report said: "The casual visitor to any one of these mills would not be impressed with the level of visible dustiness in the carding areas" —generally the dustiest part of a mill.

American textile manufacturers, seldom noted for compassion toward their workers, responded initially by denying that "the so-called disease, byssinosis," even existed. One industry spokesman grumbled that the scientists were "making a big federal case about it." Burlington Industries, biggest textile manufacturer in the world, and Cone Mills both denied that any worker in their mills had ever developed the disease.

Nevertheless, more studies in other cotton mills confirmed Dr. Bouhuys' results. Under increasing pressure from publicity generated by these reports, the industry announced in the summer of 1969 that it had already appointed a committee to study the problem. In October 1969, after seven months of supposed study, committee chairman Lewis Morris, president of Cone Mills, announced that the committee did not know what the extent of the problem was, "if any." He said the industry would sponsor "an objective and comprehensive investigation of possible respiratory diseases among cotton workers." Later the ATMI announced that the "highly respected" Industrial Health Foundation had been named to conduct the study. Thus textile manufacturers could relax. One major cotton textile manufacturer had recently commissioned a study of the ventilation at one of its mills and been advised it must spend three hundred thousand dollars to install adequate ventilation. The firm thanked the consultant and told him it had decided to await the outcome of the IHF study.

Morris had pledged to spend on the study "whatever is necessary, however much it takes." He might have added, "however long it takes," because what the industry was buying, as always, was time.

The foundation duly began to compile data, measuring workers' lung capacities, and not too surprisingly was unable to find a correlation between dust and breathlessness. Enter once again the "trace element" pursuit; foundation researchers, "struck by the very weak correlation" between dust and changes in lung capacity from the start of the working day to the end of the day, hypothesized that the breathless effect must be caused by a "pharmacologically active principle *in* the dust, but not related to the amount thereof." This agent, they theorized, was enzymes.

In June 1972 a five-member committee was convened to comment on NIOSH's recommended cotton dust standard. The members, selected to represent a variety of viewpoints, included Dr. Braun, the IHF president; Dr. Harold Imbus, medical director of Burlington Industries; George Perkel, the Textile Workers Union's research director; Dr. John Peters, a medical researcher at Harvard; and Dr. Bouhuys. What happened at the meeting illustrates the politics of industrial health. By Peters' account, three of the five, Perkel, Bouhuys, and Peters, voted for low exposure, settling on a one milligram-per-cubic-meter-of-air standard lower than NIOSH's proposed .2. Imbus of Burlington voted for .5, and Braun suggested that the committee hold off from recommending any standard at all. "He said, 'Well, hold off, we don't have enough information,'" said Dr. Peters. "'While the ATMI continues to fund me, I'll have the answer shortly.' He claimed he was on the verge of a breakthrough. But with Jim Merchant's stuff at Duke, I felt we had good information."

Dr. James Merchant told me that on the basis of his studies that there was indeed a "very, very strong linear relationship" between dust and disease. So strong, in fact, that he said he could accurately predict the prevalence of byssinosis according to the amount of cotton dust the workers are exposed to. He said the .2

standard proposed by NIOSH can be expected to produce lung disease in twelve percent of workers exposed, and that the .5 proposed by Burlington Industries would result in twenty-five percent incidence of byssinosis. Even at the .1 level, he said, "about two percent will be severely affected, by our own figures." He said .1 should be the goal, even though the incidence is still two percent, because a part of the standard would include medical surveillance which hopefully would pick up and protect that two percent.

Meanwhile, textile manufacturers applied for a variance from the present cotton dust standard of one milligram per cubic meter of air. Citing the IHF research, the ATMI asked to be exempted until the foundation could complete its investigation of enzymes as the causative agent—which should happen "within the next year and a half," said the September 1972 petition. Arguing that it was technically unfeasible to bring dust levels down, the ATMI said it "may not be necessary" anyway because the IHF research suggested that if the enzymes could be eliminated, "the dust will cause little problem as long as it is maintained below the TLV for nuisance dusts [which is *ten* milligrams per cubic meter] . . . Reduction of dust will likely lead to nothing more than the decrease of a nuisance element. Even a small amount is capable of carrying enzymes which may lead to byssinosis for some employees. With the proper amount of time, this industry intends to eliminate the cause of byssinosis and reduce cotton dust in working areas." At this writing, there had been no decision on the variance. Should it be granted, it would allow dust levels to remain at present levels at mills owned by fifteen textile companies, including Cannon Mills, Fieldcrest, Cone Mills, and Burlington Industries, representing forty percent of the industry by volume.

Merchant is unimpressed with the IHF research. "I don't know anybody considered much of an expert that is very convinced they have anything," he said, "or if they do, how applicable to control it is." He said it was hard to comment on the research because it hadn't been fully published. But, he pointed out, "One, they haven't shown any way of how these could be eliminated. I don't believe you can wait [while they make] that sort of argument. That's like saying, 'Let's wait till we find all the mechanisms for cancer before we start doing anything.' It's that sort of logic that gets people in trouble."

It is hard to believe that a research foundation so widely discredited by independent researchers can continue to flourish. But the Industrial Health Foundation will no doubt find a market for its dubious services until industry is made to deal forthrightly with the problem of industrially caused disease. Or until some IHF researcher finds a cure for "the inexorable aging process which begins at birth."

MINING

"As Safe as Any of Them"

May 2, 1972, was a chilly clear day in the mountains of northern Idaho. At 6 A.M., just good daylight at the Sunshine silver mine in Big Creek Canyon, Robert McCoy, a timber repairman, turned his pickup truck into the mine parking lot and headed for the dry house to change clothes. It would be an hour yet before the day shift crew started down, but he liked to get to the mine early. McCoy didn't like to rush, and it could take him half an hour to pull on his T-shirt, overalls, wool socks, and steel-toed rubber boots, his helmet and belt and light and battery pack, and maybe a denim jacket to wear in the breezy shaft air going down. That gave him plenty of time to discuss things with the other men —what happened up town the night before, women, fishing, hunting.

After McCoy finished changing, he walked up to the portal and poured himself half a cup of coffee from his lunch bucket—his "emblem of ignorance," he called it. Gaunt-faced and slender at fifty-six years old, he had been mining for thirty years, almost fifteen of them at Sunshine. He looked across the canyon to the ridge above the Crescent mine. Early every spring, he knew, you could see elk there in the early morning, just below the snow line, some days as many as ten or twelve head. The day before, he had spotted three, but this morning he couldn't see any.

By seven McCoy and the other men were assembled at the Jewell shaft, ready to go down. One hundred and seventy-three men. The elevator in the Jewell shaft, the "skip" or "cage," could carry forty-eight men at a time, and it took twenty minutes to lower the whole crew to the 3700-foot level, where they would board a train that would carry them back through a mile-long tunnel, or "drift," to the No. 10 shaft.

The morning started out easy in the mine. The men were relaxed, no one hurrying. At the No. 10 shaft they had time to talk some more while the skip tender finished his coffee. Then they boarded the cage for the final descent—4200 feet, 5000, 5200, 5400, 5600, their helmet lights flashing down the blurred rock sides of the shaft as they hurtled through the black, dropping thirty feet a second. There were two hoists in the No. 10 shaft, the "chippy hoist" on the 3700-foot level which hauled the men, as many as forty-eight at a time, and the double-drum hoist on 3100, a thousand-horsepower monster machine, newly installed, tricky to operate. It was used to haul muck—ore and rock—and it was also equipped with a twelve-man cage.

Operating the double-drum was Ira Sliger's job, although some days, such as this one, he had a partner to assist him. Sliger, sixty years old, looked, as he liked to describe himself, "big enough to eat oats and pull a plow," but forty-four years of metal mining had left their mark. He had had one lung removed, and the one that was left was scarred by emphysema—"dust on the lungs," he called it. All morning long Sliger and his partner, Bob Scanlan, sat in the control booth in the cavernous underground hoistroom, hauling muck buckets up and down the shaft according to bell signals from the cager who supervised the muck-loading a half mile below.

Just after noon Sliger got a phone call from a shaft crew on the 4400 level. The crew had smelled smoke in the shaft. They had signaled for the 3700 chippy hoist to come get them, but when it didn't come and no one answered in the hoistroom, they called Sliger to ask what was wrong with the chippy hoist. They didn't mention the smoke. Sliger figured the signal system must have gone out. It had failed before in the last week, and he wasn't surprised that it had apparently failed again. Underground miners keep their sanity by not worrying too much, and Sliger wasn't worried. He turned back to his controls. But immediately there was another call, this from his boss, Gene Johnson, on 3700.

"Where's your cager at?" asked Johnson. "Get him up here as soon as you can."

"What's the trouble, Gene?" asked Scanlan, overhearing.

"There's a fire down here."

Besides being the main travelway from the No. 10 shaft to the Jewell shaft, the 3700 level also housed the underground foremen's office, called the Blue Room, and the maintenance shops —the pipe shop, machine shop, drill shop, electric shop, and the warehouse. About 11:35 A.M., shortly after they finished lunch, two miners stepped out of the electric shop into the drift and smelled smoke. They yelled, "Fire!" Thirty feet down the drift in the Blue Room, foremen Harvey Dionne and Gene Johnson heard the cry, grabbed their helmets and battery packs, and ran out into the tunnel.

The two foremen began looking for the fire, following the smoke 800 feet west toward the Jewell shaft until they reached the 910 raise, a vertical shaft which went up 300 feet through old, worked-out portions of the mine. There the smoke seemed the heaviest, but they couldn't tell where it was coming from. Dionne crawled up onto the timber supports, and from there he could see smoke pouring out of the raise. By his account, he and Johnson talked briefly and decided to evacuate the mine, Johnson starting back to No. 10 shaft to give the evacuation orders and Dionne and two other men heading for the Jewell shaft to close the fire door.

In the shops on 3700, the smoke was light at first, then it pounced with force, a ball of smoke, rolling along the drift, looking to the surprised miners like a tornado, or like fog rolling down a mountainside. Hap Fowler was working in the warehouse when the smoke hit. He went back to the machine shop and turned on a compressed air line. Compressed air is used to power the drills used underground, and in such emergencies miners turn to this air supply, which is pumped down from the surface independently of the normal ventilation system. Fowler had been mining since 1929, and he had a pretty good idea how to take care of himself underground. He sat there with another miner about fifteen minutes and then decided it was time to get out. Most of the men were heading for the nearby No. 10 shaft station, away from the smoke. But Fowler figured he could make it out faster at the Jewell shaft, and he headed that way, wearing a self-rescuer, a compact breathing device that chemically converts carbon monoxide into carbon dioxide. A few hundred feet out, he ran into a blinding wall of smoke. Somewhere in the smoke he crossed paths with Pat Hobson and Jim Bush. Bush was looking for his brother.

"Have you seen Bob?" he called out.

"Yes," Fowler said, "he's still back there, if he didn't make it on the skip." He kept going until finally, at the No. 5 shaft, about halfway to the Jewell, he hit fresh air. He stumbled and fell to the ground, exhausted, and lay there a few minutes before going on.

The Jewell shaft was where air always entered the mine—and then was sucked east through the 3100 and 3700 drifts and down the No. 10 shaft. East of the No. 10 shaft, fans forced the air up through the workings and out the Silver Summit emergency exit on the 3100 level. The fire was in the worst place—on the *intake* air side of the mine—and thus it was feeding deadly smoke and carbon monoxide gas into the main airways, contaminating air which circulated through almost the entire mine, rendering the Silver Summit exit useless. There was no escape—except back through the smoke, to the Jewell shaft.

When Fowler reached the Jewell shaft, he found several men already gathered at the station. Jim Bush staggered out of the tunnel seconds later and fell to the ground. Fowler ran to him, but Bush said he was okay. Bush had tried to carry his brother and Pat Hobson when they had begun to collapse, he told Fowler, but he hadn't been able to do it. They were still back there. Paul Johnson and Roberto Diaz were sitting nearby on a motor car. "Let's go back and see if we can get them," Johnson said.

They took the motor car, disappearing into the smoke. They were gone a long time—too long, Fowler thought. He and Bush tried to re-enter the tunnel, but the smoke pushed them back. "Then Ron and Jasper, the two motormen, said they'd go back and see what they could see," Fowler recalled later. "Ron came out, he was awful sick. He was the one found Paul Johnson. Looked like they'd wrecked. Said they were all piled up. Paul and Roberto—they were all dead."

When the fire had first been detected and Ira Sliger had received the urgent call from Gene Johnson to send the cage down to the 3700 level to pick up men, the chippy hoist—the one normally used to transport men—was already out of commission. Unable to see his controls for the smoke, the chippy hoistman had had to abandon his hoistroom. That left the double-drum hoist, the one with a cage for just twelve men. At that moment it was at the 5600 level pulling muck. Its cager, Byron Schulz, heard Sliger's signal: a long-short and a 3700 station call. For all Schulz knew it was a routine station call, but when Schulz brought his cage up to 3700, he found the drift filled with smoke.

With the chippy hoist out and the 3700 level blocked by smoke, and with over a hundred men still underground, the situation was critical. Taking his orders from Gene Johnson, Schulz crammed his cage with miners and took them to 3100 level, where another tunnel led to the Jewell shaft. Then he went back down for more. Greg Dionne, a pipefitter who had come up on the cage from 3700, worked with him and together they began bringing men up. The small cage made the process unbearably slow. Each round trip, with its tiny load, took precious minutes, while the deadly carbon monoxide gas and smoke spread quickly, down the shaft, through the mine.

Five thousand feet underground, almost a mile, Robert McCoy, the timber repairman who liked to start his day at 6 A.M., was at work at the No. 10 shaft station. Just after noon, he noticed smoke coming down the shaft. It kept coming, poisoning the air.

A motor crew drove back into the drift, alerting miners along the way. The alerted miners, about twenty of them, gathered at the station with McCoy, and someone handed out self-rescuers, which were kept in a box near the station. Soon the air was a blue haze. Still the cage hadn't come. The men moved back into the drift and tapped a compressed air line, turning it on full blast. But it didn't seem to help much.

They waited thirty minutes, and finally the cage arrived. The men were so weak that Schulz, the slightly built, twenty-one-year-old cager, had to push them into the cage. They squeezed on tight, as many as could, but still half the men, including McCoy, were left behind. McCoy was still feeling fairly good, he thought, and when the cage returned and he got in, he left his self-rescuer behind, in case someone coming out of the drift might need it.

At 3100 he got out of the cage and with another miner staggered a few hundred feet down the drift. Then, too weak to go on, they sat down at the side of the tunnel, feeling sick, too sick to be frightened. A man-train stopped and someone lifted them on. McCoy didn't see or hear any more. He was unconscious.

Byron Schulz brought the second load up from 5000, returning to the 3100 hoistroom at 12:44. He stumbled off the cage, collapsed on the hoistroom floor, and then, after a few minutes, rose to his feet. Through the thick sea of smoke, he saw the floor of the big room filled with the bodies of collapsed miners. Some were still gasping for air. Some—Schulz felt for pulsebeats—were dead. Only Schulz and another miner, Doug Wiederrick, were still standing. "There's nothing we can do here," Schulz said, and they started out of the hoistroom. But the smoke was thicker in the drift. Wiederrick turned back and picked up the telephone and asked topside where they could find fresh air. "The Jewell shaft," he was told. "My God! We'll never make it," he cried and slumped to the floor. Schulz bent over him and pushed a self-rescuer into his mouth, but the miner spat it out, unconscious. Schulz, alone now, struggled out of the hoistroom and began the long walk to the Jewell shaft. A thousand feet out, he met a rescue crew coming from the surface. They were wearing oxygen packs.

"They're all dead back there," Schulz gasped. "They're all dead." He begged for oxygen, and one member of the team, holding his breath, placed his own mask over Schulz's face. At the same time, Don Beehner, another crew member, also pulled off his mask to help Schulz. At that moment, apparently, deadly carbon monoxide gas reached the group, and Beehner was felled instantly. Blood gushed from his mouth and nose. Within seconds he was dead.

In the hoistroom on 3100, the smoke had begun to bother Ira Sliger soon after Gene Johnson called for help. He had broken open the padlocked box where the self-rescuers were kept, put one on and handed one to his partner, Bob Scanlan. They had closed the doors of the glass cab in which they worked, but the smoke still came in. When Johnson arrived with the cage from 3700, he told Sliger, "You better get out before we have to carry you out."

Sliger was having trouble breathing, and he asked Scanlan, "Are you okay?"

"Yeah, I can breathe good," Scanlan said. "Only thing is, the smoke is getting to my eyes."

Sliger caught the man-train out, the same train that had stopped for McCoy. By the time he reached the surface, less than ten minutes later, topside had lost contact with the hoistroom.

Thirty-one men, including Bob Scanlan, Greg Dionne, Gene Johnson, and Doug Wiedrerrick, died on the hoistroom floor. Eighty men made it out of the mine that day. Byron Schulz was the last. Of the rest, trapped below when Scanlan died and the hoist stopped, only two men survived, Ron Flory and Tom Wilkinson. They had waited at the 4800 shaft station with other miners, but as the smoke kept billowing down the shaft, they ran back into the drift and found fresh air, brought in by compressed air lines. For seven days they waited there, in the black silent mine, until they were found by rescuers. They survived on sandwiches scavenged from their dead buddies' lunch buckets and on water from condensation on an air cooler. The final death toll was ninety-one men. It was the largest disaster in the hard-rock mining industry since a 1917 mine fire in Butte, Montana, took one hundred and sixty-three lives.

The Sunshine mine is on the eastern slope of a narrow valley cut through the sparsely forested, smelter-scarred Bitterroot Mountains by Big Creek, a tributary of the Coeur d'Alene River. Owned and operated by the Sunshine Mining Company, the mine produced seven million ounces of silver in 1971. It is the biggest silver producer in the country, a huge sprawling mine with one hundred and ten miles of tunnels. Only a small section of the mine, near the

No. 10 shaft, has been worked in recent years. Schematically the mine looks like an anthill, with major drifts following the ore deposits east and west from the No. 10 shaft, the drifts running horizontally at intervals of two hundred feet, one on top of another, and connected vertically by shafts and raises which allow transportation, ventilation, and the passage of electricity, compressed air, and water. Leading off the main drifts are short, dead-end "stopes," the producing areas of the mine. The drifts follow a major fault line which runs east and west through the mountains. Most of the metals mined in the Coeur d'Alene district—silver, lead, zinc, copper, and antimony—are found in rich deposits along this fault. This twenty-by-thirty-mile section of Shoshone County produces half the silver mined annually in the United States.

Some of the Sunshine miners are "day's pay" miners, hourly workers who handle support and maintenance jobs—the motormen, hoistmen, cagers, timber repairmen, electricians, pipefitters. The others are "gyppo" or contract miners, paid according to the number of feet they drive through the rock. Gyppo mining is hard work and the most dangerous. "I've seen young fellas come in there, twenty-one years old," says a Sunshine miner. "At thirty-five, forty years, they're stoked out, we call it. They get injured more. They have broken arms and broken legs and broken backs. But it pays well." An average gyppo miner can make fifty dollars a day and the better miners can make eighty to one hundred dollars a day, compared to the thirty-five to thirty-seven dollars average for day's pay work.

Gyppo miners work a mining cycle which ends with blasting, so that each new shift finds a pile of blasted rock awaiting it in the stopes. At the start of the shift the miners, working two to a stope, "bar down," or knock down loose rock from the "ground"—the roof and sides of the blasted area. If the ground is still unstable they may brace it with rock bolts or timber. Then they wet down the rock and "muck out," removing the pay dirt with a machine

that scoops it to the nearest raise, where it falls into ore cars below
to be hauled out. The remainder of the shift is spent drilling deep
holes into the "face" of the rock. At the end of the shift the holes
are filled with dynamite and blasted.

Nearly every man who lives in the Coeur d'Alene Valley has
mined at one time or another, and many have worked as loggers,
too. Traditionally hard-rock mining was a "tramp" occupation, and
miners moved from mining camp to mining camp, following the
high-paying jobs in the rich metal mines of the west. They moved
too often to have roots, living in tiny shacks, spending their money
for pleasure—drinking, gambling, and whoring. The miners in the
Coeur d'Alene are more stationary now, but they still live in
shacks and trailers up and down the valley. The mines—the Lucky
Friday, Star, Crescent, Bunker Hill, Galena, and Sunshine—and
the towns they support are scattered along the Coeur d'Alene River
and Interstate 90, which follows the river bed across the northern
panhandle of Idaho.

Every year four or five men die in the mines of the "silver valley"
from rock falls, haulage accidents, falls down shafts. Twenty-four
men died in the district's mines from 1966 to 1970, five of them in
the Sunshine mine. Bureau of Mines statistics were not available
for 1971, but two men died that year in a fire at the Star mine, and
one died in a rock fall at the Sunshine mine. But no one expected
a tragedy the size of the Sunshine disaster. No one ever does.
"Everyone in the hard-rock mines thought this could never happen
—nothing of this magnitude," said John Parker, manager of the
Bunker Hill lead and zinc mine in Kellogg. That is the typical
miner's view too. "The fact is, this fire was totally unexpected,"
Ira Sliger said. "In my forty-four years of mining I've never seen
anything like it. It didn't smell like anything I'd ever smelled, or
look like anything I'd ever seen. It was just one terrible fire and a
terrible disaster."

Yet it did happen. And families and friends of the trapped men came in a long procession over the wooden bridge spanning Big Creek to stare at the black hole in the side of the mountain and wait for the men to come out—or be brought out.

I got to the mine late the night after the fire and stayed up studying mine gaps and talking to reporters who had been there for a while. About 5 A.M. I walked over to the entrance gate and stood on the wooden bridge that crosses Big Creek and leads to the mine shaft. Big Creek is an incredibly clear stream that rushes by so fast that it is dizzying to watch. From the bridge, I saw "family" arriving in the frosty mountain air just before dawn, and I wondered what was happening underground—whether miners were struggling still, fighting their way out, or lying dead somewhere at the end of a drift. Aboveground the only sign of the raging fire below was a steady plume of smoke rising silently from two ventilation shafts on the side of the mountain. The smoke was trapped by the steep mountainside and settled, filling the valley with the stench.

I stood on the bridge for an hour, watching the mountains change from black hulking shadows through a range of blues until the sun hit the western hilltops, lighting the valley, and all the while, the deafening rush of Big Creek, the acrid smoke stinging my eyes, and a steady stream of family crossing the bridge, starting another day's vigil. The rescue teams worked slowly, severely hampered by dense smoke and intense heat and a series of setbacks as the fire continued to burn. They did not recover the last body until ten days later.

"It was an incredible kind of freak accident," Carl Burke, a company attorney, told reporters in a parking lot press conference at the mine the day after the fire. He said mine officials believed the fire may have started from spontaneous combustion, smoldering in old timber-filled workings near the 3700-foot level of the mine, possibly for days, before suddenly the pressure of expanding gases burst through the airtight bulkheads used to seal off the worked-out drifts. Then the poisonous gases flowed swiftly into the mainstream of the ventilation system.

But Sunshine could not be faulted, he said. "We have been one of the forerunners of the mine health and safety act. What happened yesterday, when the facts are out, will show it to be a very tragic, but a freak accident." As the facts began to emerge, however, they suggested not a freak accident so much as a familiar pattern—of choosing, once too often, to favor production needs over safety precautions.

"The trouble is," Ira Sliger told me, "the whole thing is, if they had let me know about this thirty minutes earlier, most of those men would have been alive today." The evacuation order could have been given when the fire was first discovered, sometime close to 11:35 A.M. But mine foremen had first looked for the source of the smoke, delaying evacuation until 12:05 P.M. By then 3700 level was impassable and the chippy hoist out of operation.

Hap Fowler, who had been working in the 3700-level warehouse, made the same assessment as Sliger. "The reason those guys died," he said, "no foreman wanted to take the responsibility to get those men out of there early enough. It isn't here at the mine—our people. It starts at New York—the big wheels. If they'd pulled those men out, somebody could have got in a lot of trouble—not by Chase [Marvin Chase was general manager of operations at the mine as well as a director and vice president of the company]— Chase is a hell of a nice guy. He would have backed them up. But New York, the big shots back in the east. They would have really picked a bitch. All the bosses know that. If they'd pulled those men out, that would mean they would probably lose twenty-five or thirty rounds [of blasting]. That would mean they wouldn't have no muck for the next day." And that would have cost thousands of dollars.

Fowler knew about "New York" from personal experience, he said. About a year earlier he was working for Jim Atha, then safety engineer, and asked him for some self-rescuers—"the good kind."

The reason he wanted them, he said, was that "we didn't have enough. I wanted to get some aluminum ones. Those old ones rusted out too fast. Two or three levels didn't have any on them." But Atha told him he couldn't get them.

"I said, 'Why not?'

"He said, 'They cost thirty-five dollars apiece.'

"I said, 'Jesus Christ, man, thirty-five dollars ain't much.'

"He said, 'Well, I got this letter'—he pulled out this letter. It was from New York, giving him hell for spending so much."

Atha told me later he didn't recall any such letter. He didn't recall ever getting any letter from New York.

The big wheels—the directors of the Sunshine Mining Company —are part of a group of investors who seized control of the company in a 1965 proxy fight. The group was led by the late Louis Beryl, a New York insurance broker, who a few years before had participated in a takeover of U. S. Smelting, Refining and Mining Company. Beryl and his group had barely taken their seats on the Sunshine board before they began casting about for more companies to take over. In 1965 alone, the company attempted, unsuccessfully, to merge with Kerr-McGee Corporation, Independent Coal and Coke, and U. S. Industries. In 1968 the company borrowed twenty million dollars and bought Renwell Industries, a failing Pennsylvania-based electronics firm, for twelve million dollars, including the assumption of a six-million-dollar-debt. That might not seem like the best investment in the world for a small western mining company, unless of course you were a Renwell director. One Sunshine director had been, and Sunshine stockholders charged, in a suit still pending, that Irwin P. Underweiser, the chairman of the board and president of Sunshine, and another Sunshine director benefited from the purchase, which involved an exchange of stock.

In 1969 the company bid unsuccessfully for the bankrupt Canandaigua Race Track in upstate New York. Meanwhile, several still pending suits filed by stockholders charged the board members with mismanagement, diverting funds for the directors' personal benefit and increasing their salaries to exorbitant amounts. Stockholders questioned, for example, the wisdom of an investment of "more than $1,230,843" in the stock of Reading Company, of which three Sunshine directors were also board members. Reading stock has since dropped from a high of 9⅜ in 1971 to a high of 3½ in 1972 and has been running deficits of nine dollars per share. One stockbroker told me later: "It looks like, for all practical purposes, they're well on their way to bankruptcy."

Directors meetings, usually held in New York, were "all concerned with how to invest their money," said one company critic who had access to inside information. "Most of the directors showed almost no interest in the mine. It seems like a Greek tragedy that this group is associated with this mine."

It was this management, "totally dedicated to growth through internal expansion, mergers and acquisitions," as the company boasted in its 1971 annual report, that was holding its annual stockholders meeting in Coeur d'Alene on the morning of May 2. It must have been a difficult meeting for Irwin Underweiser. After several years of stockholders' charges that the directors were "ruining the company," Underweiser had to explain why Sunshine showed a $1.2 million loss for 1971, even though the mine itself made a profit. The losses, he explained vaguely, were due to a "write-off in securities"—possibly the $1.2 million Reading investment—and lower silver prices. But he had a glowing report for the first quarter of 1972; first-quarter earnings were a healthy $122,000. "All indications are that we are definitely on the upswing," he said.

Then, in the middle of the meeting, came news of the fire. In the first few days after the fire the young New York attorney looked pale and troubled. By May 8, however, he seemed sanguine again, as he viewed his company's future. Underweiser told an Associated Press reporter that in spite of the lengthy shutdown the fire would certainly cause, "we may even make a profit on the closure." He said insurance would cover costs of a shutdown of up to six months. And since the mine is the largest U.S. silver producer, the closure could cause a shortage of silver, forcing prices up as much as ten percent.

Fires are all too common in hard-rock mines, in spite of company and industry intonations that such things never happen. The history of hard-rock or non-coal mining, like the history of coal mining, is one of fires, explosions, and disasters of other sorts. An underground fire is to be feared, not so much because of the fire itself, but because it consumes oxygen and produces suffocating carbon monoxide gas. In a metal mine such as Sunshine, the rock itself will not burn. But the millions of feet of timber brought into the mine every year to be used for supporting the walls and roofs of tunnels and shafts will burn. Until a few years ago, timbers that were no longer needed were thrown into old worked-out drifts, along with anything else the miners didn't want to haul out of the mine. Then the tunnels were sealed and the timbers left to rot and decompose. "Once they are bulkheaded off, you assume they're safe," Marvin Chase said, but he added, "It's a worry always to everyone that someday something will happen."

Company safety engineer Bob Launhardt said the company was as prepared as it could have been for a fire. He said he had "studied quite extensively in the areas of fire prevention," and found that "all major fires in hard-rock mines had been either in the intake air shaft or surface buildings. Never before in hard-rock mining had there been a major fire in other than those places."

Bureau of Mines records contradict Launhardt. Although shaft fires are more often deadly because the shafts are thickly timbered and flames can spread quickly, consuming the entire shaft and contaminating the air supply, fires have started in other areas of the mine and resulted in major disasters. Launhardt said Sunshine had focused its fire plans on prevention of shaft fires, installing concrete doors in the shaft that would be closed when carbon monoxide detectors were activated. "If it had been in the shaft, this fire would not have got back in the mine," he said. With that protection, and "having a second escapeway," he said, "I felt we were one of the best prepared mines for a fire."

Perhaps so, but clearly not the best prepared for the type of fire that has often enough hit the Sunshine mine in the past. The Bureau of Mines has recorded at least three major fires at the mine prior to the May 2 disaster, none of them shaft fires. In 1945 a fire, apparently caused by an electrical short circuit, raged for weeks before it was finally extinguished by flooding the mine. In 1967 another fire started when two miners blasted out timbers to remove them from a raise where they were working. After blasting they left for lunch and returned to find the timbers burning. The fire filled the drift with thick smoke, and though the fire was not subdued for several hours, the mine was not evacuated.

Another electrical fire in April 1971 burned along a power cable in a drift for eight feet, then spread into nearby timber. William Spear, an Idaho mine inspector, investigated the fire the next day. He discovered that the mine's fire alarm system—which warns the miners by releasing a stench gas into the ventilation system—had not been used. It was out of order. Spear remarked in the mild manner in which state and federal investigators are prone to express themselves that "most of the people that were asked said the matter was handled very well, but some thought an evacuation of the mine would have been in better judgment." Privately, Spear thought so too, and so did local United Steelworkers president Lavern Melton, who protested to management. "The company was defensive," Melton said. "They said they told the men to stand by and then went to see how serious the fire was. They said it would be awful silly to take all those men out and lose all that production for nothing."

Bob Launhardt studied theology in college and received his safety instruction from the National Safety Council, an industry-controlled organization. Miners call him "cooperative" and "conscientious" but say he is "a little bit too much 'company.' "

"The basic thing in accident prevention," Launhardt said, "is the ability to motivate people, to motivate management. Eighty-eight percent of all accidents are the result of human oversight and ten percent are the result of physical failure. Now it could be oversight on the part of management." An enlightened attitude, but it's not reflected in company practice. Sunshine's prime safety strategy has been to offer prizes to each crew for so many man-hours worked without a lost-time accident. "They're real nice prizes," said one miner, "electric fry pans, nice sleeping bags, electric can openers." Young miners thought the prizes were just so much bullshit, but old miners with families liked them. The prizes didn't discourage accidents, though. Launhardt candidly explained that the "crew incentive program" was designed "to encourage people not to stay out when they could be working." And it worked. "If a man can come out the next day," Hap Fowler said, "he'll come to work and his partner will do the work so they won't have the lost time. You get a lot of accidents, especially when you're driving new ground. It's contract, gyppo. They get too busy making money to be safe. If they'd just cut that contract out—but then the men wouldn't work."

In spite of the prizes, injury and fatality rates at the Sunshine mine have, since 1960, consistently exceeded national averages. Injury rates have been more than twice as high as the national average for metal mines, and rising—from 61.74 injuries per million man-hours in 1960 to 126.49 in 1971. The actual number of disabling injuries (or lost-time accidents) rose from 43 a year in 1965 to 133 in 1971. More than one every three days. It is hard to learn exactly what has caused the rate to go up, but it may help to know that from 1966 to 1971, in the years since contract mining was introduced, the injury rate averaged 100.33 injuries per million man-hours, up from 86.60 for the previous six years.

Launhardt blamed the mine's poor record on a rapid labor turnover and unstable rock conditions in the Coeur d'Alene. A more likely reason, proffered by some miners, was that top management just didn't give a damn about safety. "A safety engineer in the Sunshine mine has got about as much say as a mucker since about 1960," Hap Fowler said. "Since ol' Charley Angle [a former mine superintendent] left in 1960, our safety began to slow up. He really believed in safety and wanted men to eat it, live it, and think it. We were preached safety twenty-four hours a day. You'd get canned quicker for an unsafe mine practice than for missing a round or getting drunk and laying off. The guys talked safety among themselves. And the bosses talked safety. You don't see that anymore. Since he left it's not that way. Get that muck out, or else. Of course a boss likes to get muck. That's how he gets his reputation.

"The safety man today doesn't get any cooperation from the bosses, and he just hates to say anything and go over their heads. Still, the safety guy is the fall guy if anything goes wrong. He should have the power to tramp [fire] a foreman or anything else."

At times, it got to be too much even for the company safety engineer to take. Paul Johnson, who died in the May 2 fire, had been safety engineer in 1968 but had quit after one year. His widow said he resigned because "in general he just felt like he wasn't able to go ahead and do the things he'd like to do. He didn't feel like he had the backing." She said he was "a cautious person, always safety-minded. I often said, 'Be careful.' He'd say, 'I think of that first,' because, you know, he wanted to come back home."

Launhardt reportedly felt the same pressures. Lavern Melton, president of the Steelworkers Local, said that more than once when he went to the safety man with complaints, Launhardt responded, "Well, you know the problems that anyone in this job has—the limitations that he has." It didn't matter how conscientiously he went about his job. Without management support he was powerless and, as a result, ineffective. The supervisors knew it and the men knew it. And the accident rates showed it. Then came May 2.

After the fire Launhardt admitted to reporters that the company had never held fire drills, or provided safety meetings of any kind for miners. If the men were concerned about the possibility of a fire, they could read the safety manual which noted briefly, "This mine is equipped with a stench warning system. Inquire of your supervisor as to the course of action you should follow upon a fire alert." Nor were they taught how to use the self-rescuers. Explained Launhardt: "They would have to be retrained every six months or they would forget how to use them."

Astonishing as these revelations were the big surprise was yet to come. On May 7 the company announced that not eighty-two men were missing as they had previously told reporters, but ninety-three. The first number had been a guess; since the mine kept no surface record of who was underground, the best they could do was count the number of miners' lamps missing from the dry house. The only records had been kept by the shift bosses, and most of them were still underground, among the missing. The union had protested against this practice a year before, concerned that men might have an accident and not be missed, but reportedly management replied, "What we've been doing up to now has worked fine."

I left the valley in May and returned in July to do more reporting. I found the area still greatly affected by the disaster. It was hard to find anyone in the valley who hadn't lost someone in the fire—a relative, a buddy, or a friend. "It's a big family valley," Robert McCoy said. "Everybody knows everybody, 'cause they all live down the valley and they come to the same places. You can't imagine how many people were affected by this, as close friends and as many relatives as there are up and down this valley."

In addition to his work in the mines, McCoy, with his wife and son, ran a beer parlor in Pritchard, a town with eight people and four bars. In the valley, bars serve as social centers, the way churches do in some communities. The McCoys' bar, named Sag's Place, after the former owner, was a popular gathering spot. On a Friday night I dropped by Sag's Place. Three or four couples were there. A. E. Sagdahl, a county commissioner and former owner of the bar, was there too, campaigning hard for re-election. A man dressed in dirty work clothes and a red felt hat and carrying a banjo walked in, none too steadily. "Hey, Easy Money, play us a tune," someone shouted at him.

"Hah," said a gruff old fellow at the bar, a retired miner, "he can't even play the thing."

Three young men walked in and I asked the old man if they were miners. He looked them over shortly.

"Those are loggers."

"How can you tell?"

"They dress different. Usually got a good pair of boots. Trousers are shorter and they wear suspenders. They walk different, too." But it was all the same, he allowed. The men go back and forth, logging awhile, then mining again. "When wintertime comes, they all go underground, like dogs."

I approached them anyway. One with suspenders said he was Steve Johnson; he said he was indeed logging, though he had worked in the Galena mine for a while. He liked logging better. "When I was working in the Galena, there was six guys I knew that got killed after I left. You make twice as much logging, and you don't have to worry about something falling on top of you." I didn't quite understand the logic of that, but I let it pass. His companions were Terry Rice and Mick McCoy, Robert McCoy's nephew. They had also worked in the mines. "There's always two or three that get killed in a year," said Rice. "It doesn't do any good to write about it. It's not superstition, either—it's always two or three at a time—boom, boom, boom. You grow up, you know, thinking that guys get killed in the mines. It's something that happens."

I am a sentimentalist, and would have miners gathering in barrooms to sing "Sixteen Tons" and "Dark as a Dungeon," though they never do, and I thought of the song written about a silver miner killed in Comstock, Nevada:

Only a miner killed—oh! is that all?
One of the timbers caved, great was the fall,
Crushing another one shaped like his God.
Only a miner lad—under the sod . . .

It's something that happens, it's something that they talk about, that miners, workers everywhere talk about, but rarely does it worry them enough to leave the job. It's something that happens, yes—to the other guy. And the higher the risk, often, the higher the rewards in money and prestige. Miners think of themselves as a "different breed," or even "a different race" of people. They have their own code of behavior, their own rules, and in the remote silver valley mountain towns they are comfortable. They like the lack of pretense—some would call it a lack of sophistication—and the freedom that it brings.

"Nobody asks you how many years you had in school or even if you've been to college," said Johnson.

Rice nodded. "If you want to get ahead here, you get ahead. I like it because everybody's so carefree. Most jobs are simple. Nothing to detain your mind, so you're free to let it wander. There's a couple of guys in the Star, take a couple of joints down with them. Get stoned. It would be a kind of confined trip!" He told about a couple of guys in one mine who built themselves a hideaway above the timbers in a drift and installed a hot plate and a stereo. Then one day mine officials were taking some Sunshine officials on a tour and as they came down the drift it was filled with music. Rice whooped with delight. "This guy was up there cooking steaks and listening to 'Sergeant Pepper's Lonely Hearts Club Band!' "

The three lived six miles down the road in a tiny trailer they rented for forty-five dollars a month. They were young and boisterous and by local standards long-haired, and sometimes the older guys called them "or-ang-a-tangs." But nobody minded if they played hard. "They earn it," one older miner said. "They work hard."

The valley is in the middle of a huge national forest area, and hunting and fishing and camping are universally popular. Almost everyone has a camper or at least a pickup truck. The winters are long, and when spring comes, people head for the backwoods. Stores close, the theater in Kellogg closes. The bars stay open. Most people can't imagine living anywhere else. But after the fire some families moved out—especially, people report, young miners. Steve Johnson knew a guy with a wife and two children. "He just packed up and moved to Coeur d'Alene. He said, 'I've had enough.' I think more would leave, if they could." Tony Sabala, a fifty-four-year-old pipefitter at the Sunshine mine, was shaken by the disaster—his partner, Greg Dionne, was killed—but he planned to go back to work. He said: "I'll stick with it until I retire—I wish I was old enough to retire now. I've got a motorcycle I ride the hell out of. I do a little fishing, catch small trout. I like football games, basketball. I like it here. I've lived here forty-nine years and I'll die here."

The older miners were nearing retirement and they were going to go back, but just to put the time in so they could start drawing their hard-earned pensions. Ira Sliger was back at work in July, running the surface hoist for fire-fighting and repair crews. He sat back and spread his arms across the top of the couch in his tiny three-room house in Pinehurst. His wife sat in an easy chair nearby, half-listening to the conversation and half-watching the Saturday morning cartoons on the color television set across the room.

"You know, we like it here in this place," he said. He surveyed the room in a single glance. "Oh, the house don't amount to much, but it's good enough for me and Mom. I've put in a lifetime in these mines. I think I've put in pretty near enough." He worried that with all the emphasis on fire safety now, other problems might be neglected. "Oh, there's so many things that can happen. There can be mechanical failures and cave-ins. They have what they call in these mines air blasts, nothing more than a miniature earthquake. I've had them shake that hoistroom time after time. Them things can scare your socks off you. There've been a lot of accidents out there, believe me there have." But he added, Sunshine is "as safe as any of them," and the ventilation system is "as good as you'll find . . . Really, I don't see anyone too hard against the company—they have a good contract, a good vacation plan, good pension plan—that's just the reason. If I take my vacation next month I'll have thirty days' paid vacation." As soon as you have a week built up you can take off, he explained, and many of the miners like to get in some fishing in the summer and hunting in the fall. The only thing he regretted, said Sliger, who was raised in the "Tri-state" mining region of Missouri, Kansas, and Oklahoma, is that "catfish is the only fish I really like and you can't catch them here."

Hap Fowler is cut from the mold of the legendary "tramp miner," moving from job to job most of his life. He started mining gold in the California "Mother Lode" country during the Depression when there wasn't much else to do. Then Roosevelt raised the price of gold, he said, and "I just stayed with the Mother Lode. I've always made good money. Used to run from ten thousand to fourteen thousand dollars a year. And then you could travel, too. We made every major mining camp in the West. Now we've kind of settled down. Got a year and a half to go and then we're leaving." He and his wife planned to retire to the central western part of the state on the Clear Water River. The couple took turns talking—two people who knew each other so well that the thoughts of each fit comfortably against the thoughts of the other.

"Sell the shack and get us a trailer home," said Fowler.

"Go south in the winter, north in the summer," she said.

"Prospect."

"He's going to get one of those metal detectors."

"Everybody round us got killed here in the fire."

"Nine of them."

"And we want to leave."

Fowler lamented for the good old days in the mining camps when gambling was legal. "Any time you take gambling out of a mining camp," he said with disgust, "you ruin it. It makes a sheepherder town out of it. It just folds up like Jesus come to town. Any mining camp that don't have gambling, you just as well forget it. Miners have been gamblers all their lives. Used to be five saloons and ten gambling joints to every decent place in town. And cat houses all over the place. Nevada's the only state that still has it, and that's a good state." He had more to say about that, but the Fowlers were leaving early the next morning to do some prospecting over the weekend and he had packing to do.

In the aftermath of the fire, Sunshine Mining Company officials maintained that the Sunshine mine was no worse than any other mine, and possibly better than most. They were fond of pointing out that the disaster was not attributed to the violation of any laws. The federal Metal and Nonmetallic Mine Safety Act of 1966 did not require fire drills, or specific evacuation procedures, or underground oxygen supplies for hoistmen, or even self-rescuers. By providing self-rescuers underground, they pointed out, Sunshine was far ahead of many hard-rock mines. All true enough.

They neglect to mention, however, the company's own part in shaping the absurdly lenient mine safety laws that allow them to look "good" with the blood of ninety-one men on their hands. During Congressional hearings in 1965 on the mine safety act, H. B. Johnson, then manager of the Sunshine mine, wrote to oppose passage of the act as "an unnecessary imposition of federal regulation." "The industry is best prepared to meet the problems in this area through its own efforts," he argued.

The legislation—the first to provide (or at least promise) protection for non-coal miners—was passed and signed into law in 1966 during the administration of Lyndon Johnson. It showed the scars of heavy industry opposition, notably from the American

Mining Congress. The AMC is a powerful industry lobbying group, well financed, hard-nosed, and effective. It represents the interests of the likes of Consolidation Coal (subsidiary of Continental Oil), Kennecott Copper, Anaconda, American Smelting and Refining, American Metal Climax, Bethlehem Steel—in short—almost every American mining interest, hard-rock and coal.

The act entrusted enforcement powers to the Interior Department's Bureau of Mines—the next closest thing to allowing the industry to police itself. The Department of the Interior, as "custodian of the nation's natural resources," attracted special interests like flies to carrion. Its top offices were filled partially from the ranks of the mining industry and partly by political appointees. Dedicated to a philosophy of "cooperation" with industry, its officials believed in enforcing safety regulations as conservatively as possible—or preferably, not at all. The new safety law fit well with the Bureau's philosophy. It lacked provision for fines, narrowly limiting enforcement to the power to withdraw workers in case of "imminent danger" or for failure to abate a violation. Standards were to be developed by an advisory committee appointed by the Secretary of the Interior, then Stewart Udall. The early advisory committees—actually, there were three committees, because Frank Memmot, the Bureau official in charge of the act and a former American Mining Congress member, had a lot of friends to give appointments to—delayed promulgation of standards until 1969. Under the law the standards would not go into effect for yet another year. And most of the regulations were "advisory," rather than "mandatory," and thus lacked even the feeble weight of the law.

At this writing the committee, appointed in 1970 by then Secretary Walter Hickel, was chaired by James Boyd, chairman of the board of the Copper Range Company and an American Mining Congress spokesman. The committee met about four times a year but like its predecessor seldom came to any conclusions. "If [a proposed standard] isn't unanimous, it's tabled for further discussion," said Dr. Julian Feiss, executive secretary for the committee. Another committee member was Gordon Miner, vice-president and director of the Hecla Mining Company. Hecla owned rights to thirty-three percent of the Sunshine mine's production and was Sunshine's largest shareholder, holding almost four percent of the company's widely held stock.

The way that Interior officials saw things, the only problems at the Bureau of Mines were "image" problems. And of those there were plenty. The Bureau was charged with the enforcement of the Coal Mine Health and Safety Act of 1969, passed after the Farmington, West Virginia, coal mine disaster in 1968 which killed seventy-eight men. Since the law's enactment, the Bureau has repeatedly drawn criticism for its poor enforcement record, which a Government Accounting Office report in 1971 described concisely as "extremely lenient, confusing, uncertain and inequitable." Well aware that the Sunshine disaster could present yet more embarrassing "image" problems, Department officials launched an all-out propaganda campaign. Hours after the disaster struck, Hollis Dole, assistant secretary for mineral resources and often spokesman for Interior at Congressional inquiries on mine safety, ordered an embargo on information concerning the fire from the Bureau's Office of Mineral Information, and later ordered a slowdown on the release of all information on mine safety.

Dole later told a sympathetic gathering of mining engineers at a Colorado mining meeting: "In a climate of growing hostility to business in general and to the extractive industries in particular, disasters such as occurred at the Kellogg site are grist for the mills of opportunists and mischiefmakers whose real objective is subversion, not safety." Dole was a former member of the board of the American Mining Congress. In 1965, as an Oregon state geologist, he opposed passage of the mine safety law. "There's no such thing as legislating safety," he said.

He was reportedly responsible for the appointment of Bureau of Mines director Elburt Osborn, a former vice-president for research at Pennsylvania State University. Osborn was said to have told friends that he had no professional interest in mine safety and agreed to take the Bureau job only on the condition that he would not have to concern himself with such matters. He appeared at the Sunshine mine reluctantly, ordered to go. Once there, he discussed the disaster amiably with reporters. A likable man, cheerful and at ease, he looked out of place in that grim setting. Hardly out of sight of grieving families crowded near the mouth of the mine, Osborn leaned toward me and confided that the number of deaths in hard-rock mine disasters was "nothing like in the coal mines."

Death for death, the hard-rock industry in the past few years has in fact looked much like the coal industry. In 1971, 181 men died in coal mines; 162 died in hard-rock mines. The hard-rock industry employed a quarter of a million workers, about twice as many as the coal industry. But it is a varied and widely scattered industry, there is no single union as there is in coal, and the men who die in the potash fields, the salt mines and granite quarries and metal mines, die largely unnoticed, at the rate of one every two and a half days.

Secretary of the Interior Rogers C. B. Morton led the parade of department officials who flocked to the scene of the disaster. Morton expressed his sympathy to relatives of the trapped miners and commented to reporters that, "yes, sir, this has certainly been considered a safe mine." Morton, who in 1968 was chief fundraiser for the Republican Party, also conveyed President Nixon's sympathies. (Nixon sent a telegram to the mayor of Kellogg the same day, promising "the full spectrum of federal assistance"— but later refused to declare the area a major disaster zone, which would have allowed the stricken families to receive federal aid.)

Morton aide Lewis Helm was quickly dispatched to the mine to handle the department's press relations. Assigned to the Office of Communications, Helm wielded considerable power at Interior as Morton's chief publicist and troubleshooter. He has worked variously as public relations counsel for the U. S. Chamber of

Commerce, the Goldwater presidential campaign, and United Citizens for Nixon-Agnew. In Idaho, Helm was largely successful in his mission—to create the most favorable publicity possible for the Bureau of Mines. His efforts inspired news stories such as one headlined NO MINE DANGERS SEEN in the May 4 issue of the Spokane *Daily Chronicle*. The story quoted Helm, reporting: "None of the several inspections in the past two years at the Sunshine Mine has indicated any potential fire hazards, a U. S. Department of the Interior spokesman said here today." In fact, the company had been cited repeatedly for violating both federal and state fire regulations, as well as explosive, electrical, ground support, and emergency escapeway standards.

A freak accident, a safe mine, no mine dangers—it should have been no surprise, then, when Bureau of Mines deputy director Donald Schlick told a Denver *Post* reporter that the company and the Bureau were unprepared for the fire because "there has never been a metal-mine fire before. We've had small fires before in metal mines, but they were very minor and no one was hurt. No one ever expected a fire the size of the one that hit the Sunshine Mine because there's nothing to burn in a metal mine." Schlick, a former industrial engineer for Consolidation Coal, ought to have known better. Presumably he had access to Bureau statistics. In the past one hundred years, since records have been kept, major metal-mine fires have averaged one every four years, and since Schlick entered his chosen field in 1953 there have been sixty-eight reported metal-mine fires.

The man directly responsible for enforcement of the law was Stanley Jarrett, the sixty-nine-year-old assistant director for metal and nonmetal mine safety. Jarrett was considered extremely knowledgeable on mine safety. He assisted in rescue operations at the mine, directing the effort that led to the rescue of Flory and Wilkinson. But an enforcer he was not. Said a Bureau source: "He's been in the business all his life, and he is industry-oriented and doesn't even know it." Before joining the Bureau in 1969, Jarrett was safety engineer for Kennecott Copper. At an Idaho press conference he refused to comment on whether he thought the mine safety act should be strengthened. A few weeks later the Bureau announced stricter enforcement of the act, more inspectors and tougher standards, but it was an obvious attempt to undermine a Congressional move both to "clean house" and to transfer jurisdiction of hard-rock mines to the Labor Department under the relatively tougher Occupational Health and Safety Act of 1970. Whoever sanctioned the press release didn't seem to be speaking for Jarrett, who told Bureau mine inspectors at a July meeting that if death and injury rates do not improve, "we'll see legislation like you've never seen, and we don't want it."

Duffy's Café in Kellogg, Idaho, advertised Mexican food on the colorfully decorated front glass window. Inside, three scruffy teenage boys were seated at the counter, before each a large paper plate heaped with french fries, undercooked and glistening with grease. It was July, two months after the fire. The Interior Department's public hearing at the Kellogg Junior High School on the disaster had been adjourned for lunch, and Duffy's seemed a likely looking place to eat. I took a seat at the counter. A few moments later two women whom I recognized from the hearing came in and took a table against the far wall, and I joined them. Mrs. Casey Pena and Mrs. Howard Harrison both had lost their husbands in the fire. They now were angry at the news that morning that only four miners had responded to an Interior Department request asking survivors to testify. (Many had already testified, either on behalf of the company or the union, however, and most had given depositions to federal officials.)

"I think they're all afraid they'll lose their jobs," said Mrs. Harrison, a soft-featured woman in her early thirties. Her face was rounded; her eyes and hair were the same soft brown color, the long soft hair combed to fall naturally over her shoulders. She worked in the café and was dressed in slacks and a knit shirt and nylon jacket. "They'll sit around and say, 'Okie'—they called him Little Okie—'was my best friend,' or 'Casey was my best friend,' and they'll say that disaster should never have happened, but they won't even go and testify."

The grisly details of the deaths seemed to carry a special significance, the facts—real, concrete—to balance against the unthinkable horror, the immensity of ninety-one men, husbands and friends, struggling in the black against an inescapable death. The women dwelt on these details, recounting them at length. Mrs. Harrison recalled a conversation with a miner who helped carry the bodies out. "He says, 'You know, Alice, in one way you're lucky you didn't have to see their bodies.' He said it was the most horrifying, terrible thing he ever saw." The men were so unrecognizable, she said, that the crews had to ask the widows for identifying disfigurements. "Howard's got a scar on his neck from an accident in Sunshine"—she indicated, with her finger on her neck, a long gash from her ear to her shoulder—"and his foot was crushed from a cave-in at Butte. They wanted to amputate it but he wouldn't let them. It was all mangled, you know."

The women were angry too that death benefits were so low. State workmen's compensation provided a $750 burial award and a maximum $26,550 for a widow without dependent children; a maximum $35,400 with three or more children. The ninety-one men left seventy-seven widows and 181 dependent children. Three more children had been born since the disaster. There were also Social Security benefits, a $5,000 company life insurance premium, a community-sponsored $100,000 educational fund, and a $125,000 union fund, which had been divided between widows and miners laid off by the mine shutdown. Mrs. Pena thought the benefits should have been much higher—should have at least provided what the worker would have made in his lifetime. She and Mrs. Harrison were two of about fifty widows represented by a group of lawyers who were considering a possible third-party suit, probably against the manufacturers of the self-rescuers. Sunshine cannot be sued. Idaho law provided that employers could be held liable only for workmen's compensation claims.

Many of the widows were reluctant at first to join in the legal proceedings, Mrs. Pena said, but "now everybody is calling up. They're realizing that they have a complete new life to live. They feel like they should have what's coming to them." There aren't many jobs for women in mining towns. "There's office work, clerking, waitressing, tramming beer, and that's just about it," said one man.

The two widows carried with them the coroner's reports, funeral bills, military service records, and an announcement of a memorial service, listing the names of the dead. The coroner's report showed that Howard Harrison, age thirty-four, died of carbon monoxide asphyxiation in twenty to thirty seconds. Mrs. Harrison was skeptical of the findings. "You know what I wonder?" she asked. She spoke softly, almost tenderly. "Did he die that way? Did he cry? Was he scared? Did he try to climb the walls, did he try to dig out?"

I wondered aloud how the disaster had affected people in the valley. "Do you want to know what it was like before?" asked Mrs. Harrison. "This was the happiest, jolliest town that was ever on the face of the nation." She spoke fervently, her eyes sparkling.

"Do you know what it's like now? It's like a living graveyard." We stared at each other across the table. She continued: "Like, most of the miners that died, ate here. These kids were all jolly kids. Greg Dionne. Doug Wiederrick. I don't think all those men were condemned to die underground. There was one man who was supposed to retire the next week after this happened. Floyd Rais."

"No, it was next April," said Mrs. Pena.

"Was it? And William Hanna, was his partner. And Louis Goos, he'd been in a car wreck and the next day he went back to work. We not only lost our husbands, we lost a lot of good friends."

The people in the silver valley were trying to put the disaster behind them. Some even felt resentment that the inquiries and hearings continued. "It just opens up the wounds," one man said. "It was kind of sad for a while," another miner said, "but life goes on, you know." Men will keep on mining, and in time it will again be the happiest, jolliest place, but it will never be quite the same.

"The only thing that will bother me about going back to work," said Robert McCoy, "is not seeing all the old faces I've known so long. It's like you went to a party and you knew everybody in the room, and then you went into another room and it was filled with strangers. All those men, electricians, mechanics, all of them. I knew them for years."

"It hurts, it hurts," said Mrs. Mike Williams, who worked at the union hall. "I don't know if we'll ever get over it."

THE STATES

When the Law Is an Outlaw

"You don't know," said Sissy, half crying, "what a stupid girl I am. All through school hours I make mistakes . . . Today, for instance, Mr. M'Choakumchild was explaining to us about Natural Prosperity."

"National, I think it must have been," observed Louisa.

"Yes, it was.—But isn't it the same?" she timidly asked.

"You had better say National as he said so," returned Louisa, with her dry reserve.

"National Prosperity. And he said, Now, this school room is a Nation. And in this nation, there are fifty millions of money. Isn't this a prosperous nation? Girl number twenty, isn't this a prosperous nation, and an't you in a thriving state?"

"What did you say?" asked Louisa.

"Miss Louisa, I said I didn't know. I thought I couldn't know whether it was a prosperous nation or not, and whether I was in a thriving state or not, unless I knew who had got the money, and whether any of it was mine. But that had nothing to do with it. It was not in the figures at all," said Sissy, wiping her eyes.

"That was a great mistake of yours," observed Louisa.

"Yes Miss Louisa, I know it was now. Then Mr. M'Choakumchild said he would try me again. And he said, This schoolroom is an immense town, and in it there are a million inhabitants, and only five-and-twenty are starved to death in the streets in the course of a year. What is your remark on that proportion? And my remark was—for I couldn't think of a better one—that I thought it must be just as hard upon those who were starved whether the others were a million or a million million. And that was wrong too."

"Of course it was."

"Then Mr. M'Choakumchild said he would try me once more. And he said, Here are the stutterings—"

"Statistics," said Louisa.

"Yes, Miss Louisa—they always remind me of stutterings, and that's another of my mistakes—of accidents upon the sea. And I find (Mr. M'Choakumchild said) that in a given time a hundred thousand persons went to sea on long voyages, and only five hundred of them were drowned or burnt to death. What is the percentage? And I said, Miss"—here Sissy fairly sobbed as confessing with extreme contrition to her greatest error—"I said it was nothing."

"Nothing, Sissy?"

"Nothing, Miss—to the relations and friends of the people who were killed."

—from *Hard Times,* by Charles Dickens

In 1970 Adam Walinsky, a former legislative assistant to Senator Robert Kennedy ran for attorney general of New York State, and he chose to make a campaign issue of job safety. "The laws of New York State, not to speak of the elementary principles of justice, guarantee to every worker a decent, safe, and sanitary environment in which to work," Walinsky said during the campaign. "These laws are massively violated every day. Men die because of these violations. Yet in all the state of New York in 1969, there were only six fines levied against employers for maintaining unsafe conditions, ranging from a hundred dollars down to fifteen. In New York State, an employer can subject his workers to crippling injury for less than it costs him to park his car illegally."

Walinsky took his campaign to factories, mines, and construction sites throughout the state. Roslyn Mazer, his twenty-one-year-old press secretary, wrote about it later: "In Massena, along the Canadian border, men employed in the largest aluminum plants in the country—Alcoa and Reynolds—have fought their companies and the state for years over the issue of better working conditions. The president of the Aluminum Workers' Local 420, Ernest LaBaff, showed us a stack of letters six inches thick containing letters to everyone from the company president to the governor. In the pot rooms, where dangerous chemicals are used in the production of aluminum, the air is not properly vented. In some cases, the workers do not know the danger level of some of the chemicals they use. After a sixty-six-day strike in 1966 for better conditions, the state finally sent an engineer to take an air sample. The results were said by the company to be 'confidential—between us and the state.'

"A half-hour south of Massena is a town named Gouverneur where some of the world's largest lead and talc mines are. The talc millers and miners pass on their skills to their sons from generation to generation. I went into one of the mills and spoke to a fifty-six-year-old man who knew twenty men who had died of 'the dust' during his thirty-five-year employ. The talc itself sets in the air so thickly that if you would reach out to rub your fingers together, you could feel it . . . Diseases like emphysema and talcosis are the result."

Walinsky ran a vigorous campaign, producing state documents to support his claims. One was a letter from Robert Douglass, Governor Rockefeller's counsel. It was written to Martin Catherwood, state industrial commissioner, concerning the hiring of state factory and construction safety inspectors. "I passed the word along," wrote Douglass for the governor, "that any nominees which organized labor leaders wanted to suggest for these positions should go directly to Mr. Mattei (head of the division of industrial safety). As long as we are looking to add to these positions, we might as well make some friends in the process." Walinsky charged that the governor was using safety-inspection positions for politics, a charge that was strengthened by a *Newsday* story by reporter Jon Margolis. Margolis quoted a statement released by Douglass's office denying that the Douglass letter implied political favoritism. Margolis reported, however, that the statement failed to explain "why the department had not attempted to fill some of its seventeen factory inspector vacancies from among the fourteen men who passed the [Civil Service] test in 1968 but have not been hired."

Walinsky later released a confidential state labor department memo which revealed that of 1,228 deaths from industrial accidents in 1968, only 140 had been investigated, and that in fifty-eight of the 140 instances, "violations of the industrial code of the New York State Labor Law were identified as contributing to the accident." There were no prosecutions.

Walinsky went even further. He filed suit on behalf of workers at Linde Division of Union Carbide in Tonawanda, New York, asking that Martin Catherwood, the state industrial commissioner, be removed from office for refusing to discharge his duties as required by law, thus endangering the lives of the workers. State law requires the commissioner to "enforce all the provisions of . . . the industrial code" and ". . . cause proper inspections to be made of all matters prescribed by . . . the industrial code." The code provides that "all places to which this chapter applies shall be so constructed, equipped, arranged, operated and conducted as to provide reasonable and adequate protection to the lives, health and safety of all persons employed therein or lawfully frequenting such places." Walinsky argued that the existence of widespread lung disease at the plant's molecular sieve division and the state's failure to take action to protect workers from such disease was clear evidence of neglect of duty.

In a legal brief Walinsky argued that "Section 210 of the Labor Law was enacted so that citizens could call public officials to account—not for a few dollars, not for violations of laws against misappropriation or other mere matters of public money; but for their responsibility to protect the ordinary working man and woman against the sometimes terrible hazards of factories and mines, of sweated labor and industrial disease. It is a unique statute because the ills it seeks to prevent have a uniquely dangerous effect on the people of the state and ultimately on the very confidence of people in their government. Robert Moses, in the darkest days of the Civil Rights movement in Mississippi, once asked a friend, 'What can you do when the law is an outlaw?' It is the intent of Section 210 that the people of New York State should never have to ask that awful question; rather that they should have an answer, to call 'the law' to account. That is what is at issue here."

Walinsky obtained a court order, directing Catherwood to appear in court and show why he should not be removed from office. Before he was scheduled to appear, Catherwood resigned.

These newsworthy items received scant attention from the state's principal paper, *The New York Times*. A story on the court order forcing Catherwood's appearance in court was placed on the *Times'* entertainment page. "It wasn't just *on* the page," wrote Roslyn Mazer, "it was beneath Clive Barnes, beneath the opening of *Faust*, beneath Lili Kraus's 'display of boldness' at a piano recital, next to a report that a man named Bruccoli was to write the biography of John O'Hara, and next to three quarters of a page of theater ads. To top that, it was the smallest headline on the page."

New York is widely reputed to be one of the top five in state occupational safety and health programs. In 1972 Health Research Group, an organization affiliated with Ralph Nader, listed ten states with the most effective programs and expenditures made by each per worker covered by state law. The first five were:

New York	$4.83
California	2.23
Michigan	1.94
New Jersey	1.41
Pennsylvania	1.16

Others in the top ten were Oregon, Wisconsin, Washington, Maryland, and Massachusetts. The first five were often mentioned by authorities when I was first deciding where to do my research. Like Walinsky, I was interested in determining not how much money was spent, but how effectively it was spent, and whether the state program would prevent disease and injury. I found that, just as the New York program was unable to safeguard workers at Union Carbide and Syracuse Foundry, and the Pennsylvania program unable to protect Robert Ferdinand and other workers at Kawecki Berylco, so were the other states disappointingly deficient. After three years of research in fourteen states and the District of Columbia, I was convinced there was hardly a nickel's worth of difference from one state to the next as far as effective enforcement was concerned.

California, reputedly one of the best in protecting its work force, is the nation's most important agricultural state. In Tulare County, the second-largest producer in the state, I talked to the county's deputy agricultural commissioner, Clyde Churchill, about pesticides, the most important of the farm worker's occupational hazards.

Churchill told me that any grower or spraying company wanting to spray must have a permit, and he also must notify the county office whenever he plans to spray. "The grower can't buy pesticides without a permit," said Churchill. "We issue the permit for the season on organic phosphates [the most widely used of the pesticides]. It doesn't authorize them to use the material in any manner contrary to the label or the law, and we're out checking this all the time." Churchill said the county has a tough policy toward sprayers who violate the law. "We're here to protect people," he said. "We had a case last spring where a pesticide was sprayed on cars—we gave the sprayer a forty-dollar fine and two years' probation. Word like this travels fast.

"Even if it's the biggest sprayer in the county—no job means that much to me; if they are spraying workers, we're going to bring them in." Just that morning, he said, the county commissioner himself had been the victim of pesticide spraying. A sprayer had allowed the chemical to drift onto the highway, where it had covered the commissioner's car as he drove by on his way to work. "The boss said if it had been any other citizen than him, we would take it to court, but since it's just him, we'll just give him a warning."

One wall of Churchill's office was covered by a map pinpointing the locations of the colonies of bees belonging to the county's seventy-eight commercial apiarists. With the aid of the map, Churchill said, the county could warn the beekeepers—whose hives are important for pollinating crops and are highly susceptible to pesticide poisoning—whenever someone intended to spray. I learned that the county was not so accommodating to workers, however, who might want to know whether the field they were about to enter had been recently sprayed. Organo-phosphates are related to nerve gases and at first are highly toxic, but they break down rapidly into nontoxic substances. The state had developed standards prohibiting workers from entering fields for a certain number of days after spraying, the time period varying with the type of pesticide. But since the workers had no way of knowing when and with what the fields had been sprayed, they had no way of protecting themselves.

A few years earlier the United Farmworkers had sued the agricultural commissioner in nearby Kern County, hoping to force the release of the information. The Farmworkers lost. Churchill said Tulare County saw no reason to release the information either.

Why not? I asked.

"Lots of reasons," he replied. "Trade secrets is the main reason."

But, I pointed out, he had just finished telling me that the growers called him for information on what pesticides to use. So how could it be a trade secret?

Well, he admitted, that was true enough.

What were the other reasons, then?

"Well," he said, "anyone could say, 'Oh, I'm sick, I've been exposed to Guthion.' They shouldn't be able to come in and look at everyone's personal files. It shouldn't be open for anybody to just come in and paw through."

If the county officials hadn't been so concerned about visions of wild-eyed workers pawing through their secret files, perhaps Hugo Sanchez, a farmworker and a member of the United Farmworkers Union, would not have been poisoned when the field in which he and about eighty co-workers were working was sprayed.

Sanchez's wife acted as interpreter when I interviewed him. "What happened," he said, "was that the day that they were thinning the plums, they were spraying the trees also, about four rows from where they were thinning. They were spraying the trees and all the spray was going over onto the people. The workers were smelling these fumes of the spray and everything. Then they told the supervisors from the company [Tenneco] this, and they didn't pay any attention. They said it was not harmful. But they stopped the spraying. But the next day the workers went back in and were put back to work in the same trees that had been sprayed the day before, and the trees were just full of this spray."

Sanchez and the other workers quickly began to feel the effects of the spray. "We could taste the medicine being put on the trees, and we felt our tongues getting swollen from it. People felt like they had a fever inside of them, inside their nose, in their mouth and here, in their chest."

The people continued working, said Sanchez, "because they are used to this kind of work, and of course they didn't want to lose out on a day, and they didn't think it was going to affect them that bad. And in the evening most all of the people had to go to the doctor because of how they felt."

That night the union met with company officials. "Hugo was in the committee of the union. They called the supervisors, the ranchers, and the president of the company and they had a get-together that evening . . . The president of the company assured them it was not harmful, yet the people were already sick. And they still—even like they were—because these are people who really have to be sick, almost dying, in order for them to stay home —they went back and all these trees were just full of that powder. A lot of people there, they even lost their voice for a while from this powder. Upset stomachs, they used to get. It reacted in a lot of different ways in different people. First they started with their eyes watering a lot, then their nose, then they could feel it right here in their throat. And they felt like they wanted to throw up, and they all felt weak. They were all complaining, all the while that they were working, that they just couldn't stand the smell of the medicine."

"Did the supervisors tell them to go ahead and work?" I asked.

"Yes, the president of the ranch committee [for the union] is the one that talked to the supervisors and told them that the people were feeling sick from all this medicine. The supervisors told him that the medicine was not harmful—that what had happened was just that the people were very delicate and they just liked to complain.

"Most of the people would spit up, after they swallowed the

powder—they would spit up black or dark brown. A lot of people put hankies over their nose to see if they wouldn't smell so much of the medicine, but even like that it still got into their system."

Sanchez said the incident happened in part because "the companies are mad, not only with the people, but with the union. Because before, the company was able to do what they wanted with them and they could treat them any way they wanted to; and now, what's making them mad is the people have someone to defend them. They seem to pick on the people more, now that the people are in the union, than they did before. They didn't pick on them before because the people would do every single little thing they said. If they told them to go get in a mud puddle they'd go and do it."

California provides compensation for injured workers, but Sanchez said it was hard to collect for sickness. "Yesterday, a boy was working," he said, "and he got a pain and couldn't work the rest of the day, and the company won't cover him. They don't know if he got the pain from smelling the medicine or what it was, they just told him they couldn't cover him." Even when the company will pay the claim, as for an injury, the amount is so little that "if they can drag themselves out to the field, they're going to drag themselves out to the field, because they don't want twenty-five dollars instead of a hundred."

Chuck Farnsworth, an attorney for the Farmworkers, said the state paid compensation only after eight days off work. "Usually what happens with organic phosphate injuries, the worker gets weak, faints. He will be sick for a few days. It usually takes him about five to six days to recover, then he's back at work."

The pesticides that Hugo Sanchez and his fellow workers were exposed to were Diazinon, Tedion, urea, Zorba, and zinc sulfate, according to a suit filed by the Farmworkers Union on the workers' behalf. "Diazinon is an organic phosphate with a No. 1 toxicity rating," said the suit.

Through a translator, a young woman who was a relative of

his wife's, I spoke to Juan Florez, a Mexican-American worker who was poisoned in 1969 when his foreman ordered him to prepare some chemicals. "He says he was mixing up some kind of chemicals," she said, following Florez as he spoke. "Some were yellow, and then another one, light beige, and light creamy colored. And he had to mix it. And then he had another bottle, a kind of liquid detergent kind. When he went to put that liquid in the mixer as it was turning, it blew in his face. He doesn't know what it was because he doesn't understand English, he couldn't read the directions or anything like that. He just followed the directions according to what his foreman told him."

As a result, Florez suffered permanent damage. "Doctors told him they weren't very certain whether he was going to lose his eye completely, but he can't see very well with it and his ear, he can't hear out of that. The specialist told him there's no hope for it to get any better. It got the eye nerve. If he's walking, he has to walk towards one side, because with this ear he can't hear and with this eye he can't see."

In addition, Florez's face and one arm are partially paralyzed. "He says he can lift his arm, but he can't lift it high up." When he stretches, he feels his muscles stiffen. "He used to be a healthy man, before that. Right now he's still strong, he's still big, but you know, his eye, and his mouth. Sometimes when he's eating, his food just comes out of his mouth on one side."

The translator said she used to work in the fields herself but preferred the packing house where she now worked. It was hard to imagine preferring the conditions she described.

Is it hazardous? I asked.

"Well, there were, in a week, two persons—this man cut off three and then this lady cut off the tip of her fingers. It was just in a week's period. They happen, you know, on and off, little accidents. We pack frozen foods. They put chlorine in the water to clean the food before they go into cooking, you know. A couple of weeks back the restrooms got plugged up. And so in order to get out of the mess, they dumped a whole bag of chlorine—dry chlorine—on the floor where the people were working and then it got wet. Two of the girls had to go home because it got to their stomach, they were throwing up. And everybody was crying because it was burning their eyes. But it's better than working in the fields because at least it's not in the sun."

Tulare County is a part of what is always called "the fertile" San Joaquin valley. Before I saw it, I had imagined it a sort of blooming heaven, full of succulent greenery. But though it may be fertile, it is flat and dry and unbearably hot, with nothing but economics to recommend it for human habitation.

Most farm workers in the county are stationary. They live in small towns, clusters of box-like houses in rows forming box-like blocks between straight, dusty streets. Woodlawn is such a town, and there a young doctor named Lee Mizrahi ran a clinic for approximately ten thousand people in the area. The temperature was unbelievably high when I interviewed him at the clinic, as it always is in July, and Mizrahi, less concerned with his professional image than his comfort, worked bare-chested.

I asked him about pesticide poisoning, and he said he hadn't seen "much at all" that year. "This has been the year everyone's been concerned about pesticides. But what will happen when the pressure is off? This is the way these guys [the growers] have been operating for years. When the pressure's off they go back to their own dirty tricks. Nine out of ten farmers are racist. They don't care if the workers get poisoned or not. The answer," he concluded, "is the union. Until the union is strong enough to have every farm under union contract, you're going to have problems."

Mizrahi said: "I've seen more congenital heart disease—I don't know whether it's just chance or related to pesticide exposures. I've seen ten or fifteen cases—most pediatricians who just see kids see one or two or three at the most. It might just be statistical sampling that it happened that way. But these things need to be studied."

He said he also saw "an unusually high incidence" of chronic pterygium—scarring on the white part of the eye. "I see a huge amount of eye irritation," he said. "I can always tell when the sulfur is out—I see five or ten patients a day. It will happen for a week and not after that."

It is difficult to know whether people are suffering long-term effects from the pesticides, he said. "The complaints could be attributed to any one of a number of causes, from psychiatric disturbance to flu. In the chronic exposure, the complaints could be anxiety, headaches, burning eyes, burning nose when they breathe, sore throat, ache-all-over, headache, stomach-ache. Very mild symptoms that people frequently go to a doctor for.

"I'd say that one out of ten patients—this is just a guess—but for every patient that I can say, aha, this is obviously poisoning, I probably miss ten patients who are probably symptomatic for poisoning but they can't relate it or I can't relate it. Sometimes with chronic exposures to some pesticides, we don't even know their effects. I'm in the dark about it, they're in the dark about it, even the guys using the sprays are in the dark about it.

"Does it cause a chronic disease? I can't answer that. My own medical opinion would be that we would be blind to say that since we don't see anything, therefore there isn't anything. After all, we've had the previous experiences in other fields—say, the coal miner and black lung disease; people who work closer to radiation have a higher incidence of leukemia; people who smoke have a higher incidence of lung and heart disease, not to say anything about cancer. We have just a wealth of information about people who have had chronic exposures to toxic materials of different classes over a long period of time. It seems to me what happened is, the way the economy of this country runs, the burden of proof is on the person exposed, to prove that he is being poisoned, when

I believe the burden of proof should be on anyone that's using toxic materials. He's got to prove it's *not* toxic to people exposed over a long period of time.

"There's something wrong with the system, it seems to me, when it comes to pesticides. Our method is to develop the pesticides, have more crops, more prettier things to show off, and then when we figure we've been poisoning everybody, to get all very concerned about it and run around making exploitative TV shows. They wait until the damage is done and then everybody gets concerned about it. Nobody gets concerned beforehand. The press contributes to the whole mentality that goes on. When they get hip to something, they start pointing the finger as if to say, wow, we're the good guys because we're pointing the finger at the bad guys. But they've been just as much involved in the bullshit all along as anyone else."

Dr. Thomas Milby, chief of the California occupational health program, is widely respected as one of the top professionals in the field. And even the most critical physicians, such as Dr. Mizrahi, feel he does the best he can in a difficult political situation. When I visited him, Milby extolled the virtues of the state's enforcement. "As you know," he said, "when an industry wants to register a pesticide, it has to do a lot of things. It has to do all sorts of feeding studies, acute and chronic, it has to do fertility studies, carcinogenesis, mutagenesis, teratogenesis, effectiveness studies—they have to do a great deal of studies.

"The Department of Agriculture takes a very dim view of infractions of the regulations. They are generally interested in not poisoning workers. And secondly they recognize the fact, I'm sure, that if you have one agriculturist violate these regulations, you can have a lot of people sick, and that doesn't look very good for agriculture. The regulations truly are the most comprehensive of anywhere in the country.

"You're dealing with a very powerful issue," he pointed out, "with the right to make a profit. You're dealing with the guts of life, you're dealing with business. What we've tried to do is take a balanced view."

Chuck Farnsworth, Farmworkers attorney, gave me a memo which provided some insight into the politics of taking a balanced view. The memo, dated May 14, 1971, concerned a conversation Farnsworth had had with William Serat, of the Community Studies of Pesticides Division of the State Department of Public Health. Serat told him what Farnsworth called "some interesting things."

Serat spoke of a study the state had made of organic phosphates. Based on the results of that study and other studies, Serat, according to Farnsworth's memo,

recommends the following reentry periods:

Guthion	116 days
Zolone	86 days
Ethion	14 days
Dibrom	5 days

But, Serat said, such upper limits would prevent the farmer from using the stuff effectively at all, so you have to compromise with the Department of Agriculture . . . The following are his compromise figures:

Guthion and	
Zolone	35 days
Ethion	14 days
Dibrom	4 days

Still, he told me, the Department of Agriculture is resisting even these, and rather than fight them, he is willing to accept the following although *he knows* they do not assure safety:

Guthion and	
Zolone	25 days
Ethion	14 days
Dibrom	4 days

The memo reveals the process that goes on. One has to wonder what rationalizations a person must concoct to accept a system that places so little value on human life.

In Michigan, construction safety is regulated by a licensing system for contractors somewhat similar to the permit system used in California for pesticide use. The system itself has much to recommend it except that the penalty, revocation of the license, is so severe that regulators have seemed reluctant to use it. In the seven years that Michigan's construction safety law had been in effect, from 1964 to 1970, 303 men had died on construction sites, but not one contractor had lost his license. If a contractor were to lose his license and operate anyway, a state official said, he could be fined from ten to a thousand dollars.

Norm Crenshaw, one of Michigan's construction safety inspectors, worked thirteen years as a bricklayer before he decided to go to work for the state. I spent a day with Crenshaw in Flint, Michigan, while he inspected construction sites, and from what he told me and what we saw, I developed quite a lot of respect for the man with sandy hair and a deep grandfatherly voice.

"As a bricklayer I was doing much better financially," he said, "but I could see that something needed to be done. I was on a job one time and I saw a guy killed by a cave-in. He was working in a ditch and the thing caved in on him. I've worked on some scaffolds that a squirrel would be afraid to crawl on. But when you're a laborer, you're the low man on the totem pole and you can't do much about it. They're willing to risk your life for a few bucks, and somebody has to speak for them."

Crenshaw clearly enjoyed his job, but he was filled with frustration and anger that he had so little authority to enforce the laws in which he believed so strongly. Our first stop was a street excavation job, an unscheduled stop—Crenshaw noticed the project and decided to take a look. We didn't do much more than look. Back in the car Crenshaw said: "The air compressor is much too noisy. I didn't mention it because we don't have any way of measuring it. We have no maximum noise level established—nothing to go by."

On our way to the next stop he explained some of the difficulties of his job. "Before January one, we didn't have any enforcement powers," he said. "In January that was changed. We were told, 'You guys are going to be in the enforcement business,' but when it came down to the wire, we were told, 'Don't do nothing.' " He said the state inspectors investigate all fatalities, but he'd like to investigate other accidents as well. "Now, we're only supposed to investigate accidents if we call Lansing [the state capital] first. There was a time we used our own discretion but some joker put a stop to that. This is a frustrating job, but not in the field; the frustration is from above.

"Most of the other guys feel the same way I do," he added. "I've got some contractors I'd like to lower the boom on—some of them have even killed guys. But there's no penalty that we can impose on anybody. You can commit murder on a job site and make it look like an accident, and all it's going to cost you is workmen's compensation.

"A cousin of mine got killed down here at General Motors. He and his partner were walking down the aisle. There was a dolly sitting sticking out in the aisle loaded with steel panels—very top-heavy. A guy driving a tractor came down the aisle, hit the corner, flew over and hit the post. The whole thing tipped over and crushed my cousin to death. His widow wanted to sue—but she found out all she's entitled to is workmen's compensation. She has no recourse. This is wrong."

We reached a construction site where two men were working, laying pipe in a ditch about seven feet deep. Regulations required either shoring, to keep the sides from caving in, or cutting the sides back at a wide-enough angle that they wouldn't slide down. This ditch had neither, and even as we watched the men work, the wet sandy walls were crumbling into the watery sludge at the bottom of the ditch, leaving an overhanging cliff of sand and clay. It took Crenshaw several minutes to talk the supervisor into cutting back the sides. After that had been accomplished, he turned his attention to the electrical equipment the men were operating in the watery ditch. "That's a real good way to get killed there," he told me, "if that drill shorts out." He checked the line for grounding with a pocket indicator he carried with him. The drill was not grounded. When he reported that to the supervisor, the supervisor, with great reluctance, ordered the men out of the ditch. The problem was eventually traced to a short in the truck generator to which the drill was connected.

When we got in the car, Crenshaw said, "I didn't tell them they couldn't go ahead and work. I don't have that authority. I just tell them what the hazard is, and sometimes they shut down.

"It goes on all day like this. You just look for trouble and when you find it you try to get it cleared up." He wrote a violation notice against the two contractors involved in the violations—the city of Flint, which employed the men working on the pipe, and North End Trucking and Excavation Company, which was responsible for the trench.

"I never go back to see if they're in compliance," he explained, "because if they're not, I can't do anything. They'll have less respect for me because I've failed."

At the end of the day, a couple of inspections later, he once again criticized politicians who made the policies he had to follow. "Some of them," he said, "think Michigan is the leader of the pack. I don't think we're that good. If we were all that good, there wouldn't be that many dead people."

The District of Columbia's industrial safety division impressed me as the most effective and tough-minded of all the jurisdictions I saw. Charles Greene, director of the division, told me he believed the only way to stop safety violations—which he maintained were the cause of most accidents—was through strict law enforcement.

That may not sound revolutionary, but in the industrial safety field, it is. Most state and federal safety officials will say, when asked, that education is the key to safety, that strict enforcement of the laws and fining employers only antagonizes them. Greene disagreed. "Anyone who thinks education is the key to it is being misled," he said. "We tried education under another director, and it was a complete flop. No one was interested in education. But the minute we started enforcing the law, management started getting concerned. *Then* they started asking for an educational program. Education evolves out of regulation."

Greene's ten-man inspection force operated under strict instructions. They were to cite employers for every violation. Greene regarded every violation as serious; he was contemptuous of federal procedures, under the new Occupational Health and Safety Act, which allow an inspector to ignore *de minimis* violations. "We'll take a man into court just as fast for repeated lack of toilet facilities as for not having a proper railing," he said. "What gives them the right to say what is or isn't dangerous?"

The division prosecuted all "imminent danger" violations; others were prosecuted if they were repeated violations. "Otherwise," said Greene, "we'd be in court every day." Violators faced a fine of from $100 to $600 per violation, and ninety days in jail. The number of prosecutions seemed limited mainly by the division's shortage of manpower.

Construction is the District of Columbia's major heavy industry, and in 1971 construction accidents caused twelve deaths and 7,082 reported injuries. In 1972, when I was investigating construction safety in the District, the city was well along on a $2.98 billion rapid transit system, a gigantic project. Seven miles of the ninety-eight-mile rail and subway system were under construction and the project was having trouble with safety. According to Clyde Farrar, Jr., chief inspector for the division, one out of every three men working on the subway had been injured since work began in January 1970.

Construction is hazardous at best. Each year in the United States an estimated 2,700 construction workers are killed and another 250,000 injured. Tunnel work is one of the most dangerous types of construction. Tunnel workers must contend with rock falls, dangerous natural gases, fumes produced by machinery, and dust generated by blasting and drilling. Explosives, which must be stored and handled in large quantities, carry an inherent risk. Even with careful planning, no one can predict all the potential hazards. At one Metro site, tunneling workers uncovered pockets of fuel oil which constantly burst into flames as the men worked. Apparently the oil came from leaking underground storage tanks.

Occasional major disasters call attention to the hazardous conditions. In December 1971 in Port Huron, Michigan, men were at work in a six-mile tunnel under Lake Michigan when the tunnel exploded. Newspaper accounts listed twenty-two dead, but a Bureau of Mines expert told me the men were so badly mangled that authorities weren't sure how to count the pieces. Most construction fatalities, however, come one or two at a time and never make the headlines.

The Washington Metropolitan Area Transit Authority, WMATA, the public agency authorized to plan, build, and operate the Metro system, said its injury rate was no better or worse than on any construction project. WMATA claimed a frequency rate for December 1971 of 29.6 injuries per million man-hours worked and a "severity rate" of 896 days lost per million man-hours. While the statistics were not out of line with national figures, accident rates were rising steadily. They were also inaccurate, since the statistics counted only lost-time accidents. District Employees' Compensation Deputy Commissioner Noah Walter said that in the last three months Metro contractors had reported more than a thousand injuries in which the victim stayed on the job. "We have reports of back injuries, head and knee injuries from falling materials," said Walter. He estimated that five to ten percent of the no-lost-time injuries reported to the commission were serious.

"Often," he said, "when they suffer injuries, although the injury normally would be considered serious, compensation is only seventy dollars a week, and they just can't afford to stop working. Many are from out of town and they have to send money home to their families."

Harlon Cheney, a thirty-three-year-old operating engineer, was not a part of the statistics. Although he was off work two months with a broken leg following an accident on the job, a check of the records showed that his contractor had reported only thirty-eight days of lost-time accidents to that date.

Cheney remembered the accident vividly. It was October 13, 1971. Workers were digging a shaft from which they eventually would tunnel under the Potomac River, from the Washington banks to Alexandria, Virginia. Though it was October, it was hot at the bottom of the thirty-five-foot shaft where Cheney was working. Most of the day he had been operating a back hoe, scraping out rocks and dirt and piling them so they could be removed by a crane operating at the surface. The rocks were damaging the bucket on his machine. He climbed out of his cab and knelt beside the loader to change the bucket for a bigger, stronger one.

He was not happy. In spite of his objections, the crane continued to work just above him, lifting ton-loads of rock and dirt out of the shaft. Cheney had watched the crane many times before and noticed that often the crane lifted its clamshell bucket before the jaws were closed—sometimes allowing the whole load to drop before the bucket reached the surface, rocks often falling around his machine, even bouncing off the protective cab where he sat. As Cheney worked beside his back hoe, he kept an eye on the crane boom swinging over his head, watching as it lifted the loaded bucket again and again—four times, he counted—without incident. Then, on the fifth load, as Cheney looked up, the bucket failed, releasing an avalanche of rocks directly above him. He crouched under a beam at the side of the excavation, pressing himself against the wall. Most of the rocks bounced off the beam, but one hit him on the leg, breaking it in two places.

Cheney was unfamiliar with workmen's compensation laws. Lying in his hospital bed later he thought about suing the company, but decided instead to ask the contractor, Shea-S&M-Ball, if it would pay his salary, about three hundred dollars a week, while he recuperated. "They said, 'Heck no,'" Cheney related later. "'If you'd been with us twenty-five years, we might have considered it.'" Then he consulted a lawyer, only to discover that all he was legally entitled to was workmen's compensation benefits of seventy dollars a week.

"It's not right," said Cheney, who had a wife and child to support. "It's no way to live on seventy dollars a week. It won't live anybody. You can make your groceries and car payment if you have one, but as far as your rent or anything else goes, forget it."

Metro's insurance carrier was National Loss Control Service Corporation, a division of Kemper Insurance Company. Its representative, Ken Kueper, was a man with a national reputation in construction safety. He was associated with the development of Du Pont's construction safety program, regarded as by far the best in the industry. "There's too great an exposure risk out there right now," Kueper told me. "Unless the work methods and the control of the contractor improve, I cannot foresee any insurance company continuing to underwrite the present type of exposure. We could pull out and pay the penalty for breaking our contract and we may be dollars ahead."

He said Metro officials were not enforcing the safety codes that had been written into all Metro contracts. "To have an effective safety program—which we don't have here—it's got to start with top management, and management has to insist that it is implemented." People will perform "to the degree they're caused to perform," he said. "The worker has to have that job to survive.

He'll do exactly what he's told to do. But we have the legal responsibility for that man on the job. If he isn't doing it right, don't look at him. Look at the foreman, then the superintendent, and up to the president. If they don't want it up front, you aren't going to get it down here from the worker."

Shortly after I interviewed him, Kueper told WMATA officials exactly what he thought of their safety program. A few days later, he was forced to choose between another job assignment with a cut in salary or resigning. The carrier, he explained, "doesn't want to eliminate all accidents. If the rate goes too low, they would have to adjust premiums and they would lose money, because at a lower rate the cushion of profit isn't as great. They're duping the public."

I had just completed my investigation of the Metro project when Frederick Gau, a twenty-seven-year-old engineering student from the University of Illinois, was killed. The first man to die on the project, Gau was crushed to death by a six-ton wedge of rock in a roof fall. Safety inspector Clyde Farrar was in the tunnel before the dead man could be carried out. He blamed the contractor for the death and charged them with two violations: failing to install adequate bracing for the rock and an unauthorized work practice not directly related to the accident.

WMATA released the following explanation:

> A high Metro official said today no negligence appears to have been involved in a subway construction accident yesterday that claimed the life of one man and injured two others.
>
> George Kline, senior construction engineer with De Leuw Cather and Company, general engineering consultant to the Washington Metropolitan Area Transit Authority, told Metro's regular Thursday morning board meeting that he visited the site shortly after the mishap.
>
> "There was nothing I could see that was due to any negligence. It was truly an accident," Kline declared.
>
> General manager Jackson Graham said Metro's severity safety

record is far better than the national average. He explained that this was the first construction fatality of the Metro program in the 26 months and four million man-hours of work since ground-breaking. The national figure for tunnel work, he noted, is 1.5 fatalities per million man-hours, while the rate for heavy construction work generally is about one fatality per million man-hours. In this sense, he told the board, Metro was "statistically overdue."

Rev. Jerry A. Moore, Jr., Metro vice chairman, led Metro's board and staff and spectators in a minute of silent prayer for the deceased worker.

Employers so often hide their industrial injuries and deny the existence of work-related disease that it is impossible to quote accurate statistics. And since industry and government have traditionally emphasized accidents, reporting methods are grossly inadequate to measure occupational diseases.

The most recent government statistics available in January 1974 are from a mail survey of 60,000 employers. For the last six months of 1971, among a total work force of 57 million workers (not including farm, railroad, or mine workers or government employees) the Labor Department estimated 3.1 million "recordable occupational injuries and illnesses" and 4,300 deaths. Less than five percent of the 3.1 million figure, or 133,000, were recorded occupational illnesses. "The statistics, however," stated the 1973 *President's Report on Occupational Safety and Health,* "may not reflect all occupational illnesses since some illnesses of occupational origin may not have been recognized as such." That is an understatement. It is common for employers to contest even the most obvious cases of total permanent industrial disease when workers file for workmen's compensation; it is hardly likely that they will report as occupational diseases the very cases they are denying in court.

In *The President's Report* of 1972, NIOSH estimates of occupational disease are indeed alarming: "Recent estimates indicate at least 390,000 new cases of disabling occupational disease each year. Based on limited analysis of violent/non-violent mortality in several industries . . . there may be as many as *100,000 deaths per year* from occupationally caused diseases" (emphasis added).

An additional problem which makes occupational diseases and even injuries difficult to measure is that often workers, so accustomed to the discomforts of working with irritating gases, fumes, and dusts, come to accept as normal many symptoms that would immediately have sent a white-collar worker to his doctor. Worker fears may play a part too—realistic fears that he will lose his job or be placed in a less desirable job because of his illness.

In 1971 a worker at an Allied Chemical plant in Syracuse, New York, told me that the company had pulled him out of the "mercury cell" department where he worked early that spring because of elevated levels of mercury in his urine. "I had extreme tiredness," he said, "and still do, and sore and aching muscles. I go to the bathroom more often, and I have trembling—in my hands and arms. When I was pulled out it was quite bad. Lots of times I'd be sitting at the dinner table and I couldn't hold my fork. I'd have to switch hands." He also complained of irritability, a usual symptom of mercury poisoning. "I think my wife could vouch for that—every little thing irritates me.

"They don't call it mercury poisoning, they refer to it as 'high mercury content,' " he said. "You know, the thing is, there are a lot of guys out there that have it just as bad, but they won't go to the doctor. I suppose they're afraid they are going to lose their job.

"I'm trying to find another job," he said, "but it's not easy. In October I'll have fifteen years there. That'll guarantee my pension."

Boredom on the job and "worker alienation," began to receive a lot of attention in the early 1970s, and while these are real issues, researchers and the press, unaware of the tremendous hazards in manufacturing plants, missed the point. Talk of "job enrichment" has masked the far more serious problems of disease and injury. It is often these problems that lead workers to be dissatisfied with their jobs. The worker who is wheezing and vomiting from chlorine exposure, for example, or constantly subjected to tremendous heat and noise, is likely to be a dissatisfied worker.

A 1971 report on a survey of worker attitudes * made the point clearly. The survey, which had included both blue-collar and white-collar workers, found that among the problems most important to workers were those relating principally to the general area of health and safety (including health and safety hazards, unpleasant physical working conditions, work-related illness or injury, and wage loss following a work-related illness or injury) and, secondarily, to the general area of income . . . The report went on to say: "Workers reported 'becoming ill or injured' as the labor standards problem against which they most wanted to be protected, and 'health and safety hazards' as the area involving some of their most frequent and severe difficulties. In addition, 33 percent reported working under physical conditions they felt were not necessarily hazardous but were unpleasant or irritating. Workers reporting problems with such unpleasant physical working conditions principally complained about difficulties with heat (or lack of heat), overcrowding, unclean conditions, and inadequate, antiquated, or uncomfortable furnishings. Finally, a full 13 percent of the workers interviewed reported that they had actually experienced a work-related injury or illness in the last three years. Of these, 42 percent reported that the injury had kept them off work for two weeks or more."

A 1968 survey of health hazards in five hundred Denver businesses—the survey included businesses in construction, manufac-

* "The Working Conditions Survey as a Source of Social Indicators," by Neal Q. Herrick and Robert P. Quinn, published in the *Monthly Labor Review* and reprinted in Senate on hearings on "Worker Alienation."

turing, wholesale and retail trade and selected services, each employing fewer than two hundred and fifty workers—revealed a high incidence and wide variety of hazardous exposures:

> Inspections of the plants revealed that exposures to hazardous agents and materials were common, averaging about thirty exposures per establishment. Industrial hygiene controls of these exposures were absent or inadequate in more than one-third of these situations. Almost 25 percent of the plants had hazards which the industrial hygienist adjudged serious enough to warrant immediate attention.

> The estimated population at risk was sizable. About 30 percent of the study population, or almost 43,000 workers, were employed in establishments which were rated "high priority," that is . . . serious enough to require a visit within one year . . . Multiple, mixed, and unidentified chemical exposures were frequent. All the establishments which were rated as high priority had chemical hazards; in three-fourths of these, a physical hazard was also present. In nine of ten high-priority establishments, the employees were exposed to hazardous gases, vapors, fumes, and mists. Noxious and pneumoconiosis (or lung-disease)-producing dusts were present in 42 percent of the high-priority establishments, and significant skin irritants were found in 66 percent.

The high-risk plants averaged fifty-five inadequately controlled chemical exposures for each hundred workers. Three-fourths of the employees in the high-risk plants were in manufacturing and trades. That meant that the precentage of hazardous plants would have been even higher if only those industries were included.

All in all, such statistics as there are portray a massive, hidden national agony in the American working place and in the families of the maimed and blighted and killed.

EXPLOSIVES

People Are to Use

The rocket test pit is slowly filling with water. The barefaced steel beams are spotted with rust, and marsh grass has crept between the cracks of the concrete pad encircling the pit.

The pit, used once and then abandoned, was part of a multi-million-dollar plant that the Thiokol Chemical Corporation opened in 1964, under a National Space Administration contract to develop and test solid fuel rockets. It was an ideal site—seventy-five thousand acres, mostly marsh land, inhabited by wild turkey, boar, rattlesnakes, and deer, on the southern Georgia coast at Woodbine, only a few hundred miles from Cape Canaveral. The rockets, weighing hundreds of thousands of pounds, were to be assembled in the pit, tested, carried a few yards to the dock, and then shipped south to the space center. It was an important and exciting contract for Thiokol. Years later, long after it was over, company officials still

263

would reminisce about the program, about the first 800,000-pound rocket, then the world's most powerful booster rocket, built and successfully tested in this pit. A second, even larger, was under construction when NASA canceled the contract in favor of liquid fuel rockets.

The complex, built at government expense, was too valuable to waste. In the years that followed, Thiokol scrambled to pick up new contracts, jobs suited to a rocket manufacturing facility. They were a diverse assortment: government contracts for tear gas bombs, 81-millimeter flares and trip flares for Army use in Vietnam, systemic insecticides; and various commercial products—caulking compounds, nail polish, a hair removal chemical for Revlon. The workforce dropped from five hundred in 1965 to less than two hundred by 1969. Thiokol paid minimum wage for many of the new jobs and hired primarily black workers, previously un-employed, and many on welfare. "Unemployables," the federal Manpower Administration considered them, paying Thiokol $1,695 per worker to hire sixty employees at $1.65 an hour, teach them basic English, math, and personal hygiene, and train them to work as shell assemblers. According to the terms of the contract, they would learn a variety of tasks preparing them, theoretically, for advancement—assembling, inspecting, measuring, and weighing—and they were to receive twenty hours of safety instruction as well.

Most of the Manpower workers came to Thiokol less from choice than from necessity. To qualify for welfare in Georgia an applicant must accept work if it is available. Jobs were always available at Thiokol. The work was not hard, but it was often hazardous. Turn-over was high. Not infrequently, a would-be worker would hire in, look the place over, and quit on the same day. Understandable, if the experience of a woman hired into the "C.S." tear gas depart-ment was at all typical.

"C.S. is very dangerous," she said. "If it gets to your skin it will make you sneeze, cry, itch, and leave your skin as if you were burned by fire." She worked in the department three weeks before she complained that she "couldn't stand it any longer" and was transferred to the packing department. There she was assigned to lifting sixty-pound boxes with seven other women. "We complained about the heavy boxes so much that they sent us to 137," she says. Building M-137, a part of the munitions complex, proved to be no better. In the first week there, she witnessed her first fire, a frequent occurrence in that department. "We all ran out," she remembered. "Some fell down, some were hollering, but soon the fire was put out." Then she was transferred to M-132.

On February 3, 1971, fifty-five workers were in building M-132, manufacturing trip flares to be used by the Army in Vietnam. The flares are hidden in the brush, and secured to the ground with wire. When an enemy soldier trips on the wire, the flare, filled with a magnesium powder, bursts into brilliant flame, illuminating the area and exposing the enemy's position. The flares consist, basically, of four "illuminant" pellets made of magnesium and sodium nitrate, one pellet containing extra "first-fire" ingredients to set the flare off when it is tripped, and an aluminum casing called a "candle," about the size and shape of a beer can.

Once the "illuminant" and "first-fire" ingredients had been separately mixed and cured, the powder was formed into pellets on an assembly line, then baked at 110°F in the curing room. Finally the pellets were loaded into the candles and sealed. The whole operation, except for the manufacture of the cover assemblies, was completed in building M-132. The cover assemblies were made in an adjoining trailer.

Eight women were working on the illuminant assembly line in M-132 the morning of February 3. At the top of the line were two women, cousins. The job of the first woman, the "die turn-around operator," was to lift a pellet die—a mold—from the moving conveyor belt to a wooden platform. She would hold it while her cousin, the "first-fire dispenser," scooped from a barrel a measured amount of the first fire—a highly flammable, friction-sensitive mixture of boron and barium chromate—then scraped the excess powder off the brass scoop with a plastic rod and emptied the scoop into the die. Then the first woman tapped the die gently, to level the powder, and returned it to the line. The job was simple but tedious, and demanding. To keep up with the line the women had to complete their task once every ten seconds. Sometimes they could not help falling behind. They had fallen behind this morning. About eight dies had accumulated on the wooden platform in front of them.

Suddenly, from somewhere—perhaps from under the conveyor belt—flames shot into the air around them. The fire bolted from die to die, down the line. Workers screamed, turned and ran, and as they fled, the flame looked like a brilliant ball, a "fiery hand" reaching out to grab them. The fire might have stopped there, at the assembly line, as had numerous fires before, but this time it spread to highly flammable materials stacked near the line and throughout the crowded building, racing down the corridor toward the curing room. In the curing room were thirty thousand pounds of flares in various stages of production—powder, pellets, and candles stacked together in open trays. One gigantic powder keg.

In the trailer next to M-132 were fourteen women just back from their morning break. They were in no hurry to get back to work. Josephine Walker, the compressor machine operator, had run out of powder. She sent a woman for more and sat at her machine, talking with the others. She had been at Thiokol for four months. At first she had worked in building M-137, but she had broken her hand in a fall while running from a fire in the building. Now she was on light duty, running the compressor in the trailer. White and from the north, she was not a typical Thiokol employee. Her family were all rubber workers in Akron, and she would have preferred to be there too, but for her husband. Clifford Walker had worked construction for years, and the family had moved across the country from job to job. Finally it had seemed time to settle down, raise the children. The Walkers moved to Woodbine, Georgia—his mother lived there—bought a trailer, and looked for work. There was not much work near Woodbine—just Thiokol and the pulp mill. Walker got a maintenance job at Thiokol, paying $4.65 an hour, and later his wife went to work there too at $1.65 an hour.

Spirits were high in the trailer that morning. One woman had learned a new dance Saturday night and she was demonstrating it. Then came the cry: "Fire!" Josephine Walker looked through the window toward M-132. White smoke was pouring out. "My God, yes!" she screamed.

The woman ran for the door, Claudia Jacobs first and Josephine Walker right behind her. She pushed open the door, and suddenly she felt herself swept into the air. Everywhere was brilliance—"like a million stars," she thought, suspended in space. The stars were sixty thousand ignited trip flares spewing white fire—just one will light up an airfield runway—raining all around her. She landed two hundred feet from the trailer, surrounded by fire, badly burned, stunned by concussion. Determined not to lie there and burn to death, she began crawling toward the road, screaming for her husband.

Clifford Walker was in the maintenance shop, five hundred feet away. When M-132 exploded, it blew out the end of the shop, throwing Walker sprawling across the floor. He rose, unhurt, and ran outside. Fire was everywhere. He found his wife on the road.

The explosion had been almost unbelievable. Scientists measure the force of an explosion in pounds of pressure per square inch. Sudden pressures of one-half to one pound per square inch will shatter glass. One psi will begin to injure human beings; the most fragile blood vessels will rupture, causing bleeding from the ears and nose and mouth. Three psi creates death from concussion. It was later estimated that at its center, the Thiokol explosion was from 500 to 1,000 psi.

Less than sixty seconds after the fire began on the assembly line, twenty-nine workers lay dead or fatally injured. Another fifty were wounded, some missing an arm or leg. Bodies had been thrown hundreds of feet; one observer counted fifteen bodies in the burning woods, lying where the blast had hurled them.

Rescuers started arriving within minutes. The wounded were evacuated by every available means—cars, ambulance, helicopter, and airplane. One pilot told of flying a wounded woman to a nearby hospital. "She was conscious, and kept trying to reach over the seat to me," he said. "She cried, 'Please help me,' then 'Please shoot me.' I'll never forget it."

The explosion demolished building M-132 and caused major damage to buildings more than a hundred feet away. One hundred and forty acres of timber burned. Later, people would liken the blast to that of an atomic bomb. In Kingsland, ten miles to the southwest on U.S. 17, the ground trembled, shaking buildings and rattling the window panes. Eight miles across the bay in Saint Marys, a former Thiokol worker, once union safety steward, heard the boom and frowned. "It's that gol'damned Thiokol," he thought to himself.

The next day, after the bodies had been cleared away, Thiokol officials began poking through the wreckage. They were joined by Bernard Addy, inspector from the U. S. Department of Labor's Bureau of Labor Standards. Under the Walsh-Healey Act, the Bureau was charged with regulating safety and health in factories operating under federal contract. Addy visited the plant, decided that "very little . . . could be learned from the standing remains," and left. A week later John Lankford, another Bureau inspector, visited the plant. Lankford also refrained from launching an investigation; the Department of Labor had ordered him merely to "see that the investigation Thiokol Corporation itself was carrying out was a comprehensive one." Sixteen federal and state agencies sent officials to the site of the disaster; not one conducted an investigation of the explosion.

It was to be the single largest industrial disaster in the United States during 1971, yet it got little press attention. *The New York Times,* for example, ignored the story completely. The incident was swiftly forgotten. Several months later, when I began inquiring about the explosion, even Labor Department officials in Washington had difficulty remembering it. One ventured that Thiokol was not a government contractor at the time and therefore not subject to government regulation—an explanation absurd on its face, since trip flares, having no nonmilitary use, are bought only by the government. Moreover, a call to the Pentagon revealed that of Thiokol's $101.5 million in defense contracts for 1971, $8 million had gone to the Woodbine plant, including a $1.3 million Army contract for the trip flares.

I brought this information to the attention of the Labor Department, whereupon the agency took another tack. Officially, it went this way: "Plans for follow-up under Walsh-Healey revolve about standards and regulations which may be developed from an analysis of the causes of the accident as revealed by the investigation conducted by eminently qualified experts in explosions." Unofficially, from the same Labor Department spokesman who had just drowned me in double-talk: "Frankly, and I know you're not going to quote me, Walsh-Healey obviously pertained. Frankly, it looks like a botched job."

Safety inspection reports kept by the Bureau of Labor Standards are public information under the Freedom of Information Act; the problem, I discovered, is tracking them down. Washington headquarters referred me to the Atlanta regional office; Atlanta claimed the reports were in Washington. Finally, by appealing to Joseph Loftus, a special assistant to the Secretary of Labor who is a former labor reporter for *The New York Times,* I got a fairly complete set of files; but the company report was missing. Again I queried Loftus. The report was missing, he discovered, because it had never been received. "According to the Atlanta file," Loftus wrote in a memorandum to George Guenther, head of LSB (which by this time was a new agency, the Occupational Safety and Health Administration—OSHA), "our man Lankford was to see that Thiokol did a thorough job of investigation and filed a report with us in Washington. No such report has apparently been filed and Lankford is working in Birmingham. Just wondered if you wanted to pursue."

Three weeks later Guenther forwarded the Thiokol report to Loftus, with a short memo: "Please check with me before releasing publically [sic]. GG."

The company report was an exhaustive document, a lengthy compilation of error upon error. It was carefully worded to avoid faulting management practices, but the list of errors spoke for itself. Throughout building M-132 conditions were cramped and women labored in close quarters, virtually surrounded by barrels, boxes, and bins of hazardous materials. The report confessed that in the north corridor the ten workers stationed around the ignition-pellet assembly line shared space with two drums of illuminant pellets, two metal cans of granulated illuminant, a fiber drum of granulated illuminant, and, incredibly, trays carrying 250 pounds of illuminant. All this in addition to the illuminant and first-fire dispensers actually on the line. Six more drums of illuminant pellets lined the narrow passageway leading to the cure room.

Conditions were so bad, in fact, that Melvin Cook, one of two consultants whose independent reports were attached to the company report, wrote that "good housekeeping is a cardinal safety factor. From all I was able to glean it was being flagrantly violated in the Building 132 operations at the Woodbine plant at the time of the accident." In a letter also attached, apparently written in response to a company query, he added: "The presence of large quantities of ingredients and in-process materials with many people present in an apparently crowded condition is, in my mind, poor housekeeping."

The report concluded that the fire began at the first-fire addition station on the ignition pellet assembly line, probably when an operator bumped or dropped a die, or struck the brass scoop used for loading first-fire against a die. "Major contributing factors" to the "accident," according to the report, were inadequate separation of materials, and inadequacy of the fire protection system. The report went on to recommend that Thiokol adopt a number of procedures in the future that should have been obvious to any company engaged in the manufacture of flammable materials— safe methods of handling and storage. From the conclusions and recommendations, it seems clear that the disaster could have been avoided, if Thiokol had taken the trouble to evaluate the hazards of trip flare production before the explosion.

Of four pyrotechnics experts I talked to, all expressed amazement at the huge quantities of materials in the Thiokol cure room. And not one was surprised at the resulting explosion. One chemist who had worked with sodium nitrates for some twenty years and was particularly familiar with such munitions work as navy flare and howitzer mortar shell production, was vehement in his expressions of disgust at conditions I described to him—of thirty thousand pounds of trip flare materials in one curing room. "Have you ever worked with that many pounds in one curing room?" I asked.

"No, never. This seems to be a very absurd condition," he replied. "Certainly in the building [where people are working] you should have an absolute minimum of unprocessed items. This is one of the real hazardous situations, where too much of the material accumulates." Unfortunately, such conditions are common in the industry, he added. "Probably the people are not aware of the hazards. But they should know you always have to figure an accident *may* happen and what then?" When accidents do happen, he said, it is generally because management "doesn't have any imagination" and "they want to do it as cheaply as possible."

"One consideration is always expense," agreed U. S. Navy chemist Bernard Douda, who had run tests on illuminants. Such mishaps, he said, are "not all that uncommon . . . Safety costs money, always. Some people are willing to take more risks than others, some corporations are willing to take more risks than others. There may be an urgency to produce an item by a given date. They've got a contract, there isn't time to build a new facility—these kinds of pressures are very real to a production agency."

And Thiokol seemed to have had some very real production problems. Of the first eight lots of trip flares submitted, seven—or thirty thousand flares—were rejected by the military. It wasn't until four months into production that the plant began producing efficiently. Thiokol had to finish production of trip flares early in 1971; in March the building would be needed for a new contract. But by February 3, only 400,000 of 754,000 contracted for flares had been completed, and it seemed unlikely that Thiokol would make the deadline. Said one Thiokol worker, "All you ever heard was production, production. Hardly anything about safety."

After the disaster, Thiokol officials maintained to reporters that the company was operating "well within" safety regulations laid down by the Army contracting office. And beyond that, company officials had no comment. Not that local newspapers persisted with further inquiries. R. T. Buffington, director of safety and personnel at the Woodbine plant—who when asked to explain would say only, with the air of one grievously wronged, "There is a tremendous story behind this, but the time is not right to say anything about it"—told me newspaper coverage had been "understanding." He suggested I talk to Howard Davis, publisher of the *Southeast Georgian* in Kingsland. Davis said: "The reasons for the explosion never came out. They said it was so highly technical no one could understand it."

As for operating "well within" Army safety regulations—hardly. The contract required "a system safety engineering plan which will (1) assure that maximum safety consistent with military operational requirements has been designed into the system," and (2) assure that "adequate controls over known hazards, inherent to the product, are established to protect personnel, equipment and property." Further, the contract provides that "where human performance is critical to safe operations, training requirements and lesson plans will include safety information; exams will test personnel knowledge." Safety meetings were so infrequent and turn-over so high that few workers could remember ever attending one. Nor had there been any fire drills. In fact, women said they were told not to run in case of a fire because "boss man said wasn't anything that could hurt you."

State regulations were violated as well, according to worker testimony. For example, state law requires that "all stairways, passageways, gangways and ramps shall be kept free of materials, supplies and obstructions." Workers testified that doors and pas-sages were often blocked with supplies. "I can recall many times the two largest doors in the building being locked," said one worker. "The passageway seldom stayed clear for two people to pass side by side. A small person had to turn sideways to get by." Workers also complained of powder on the floor. "First-fire, which is very dangerous and could catch at the scrape of the feet, was always on the oven floor."

Though workers for the most part would not talk to a stranger about the conditions in the plant, the consensus of opinion among those who would was that Thiokol was a hazardous place to work and that the company made little effort to make it less so. Said a former union steward, "They've had nothing but accidents at Thiokol. At one time I went out to the hospital, and they had about twelve people out there at once. I'll bet I wrote three hun-dred grievances just on safety. Half the fire extinguishers didn't work; safety regulations were never strictly enforced. Everyone out there will tell you they weren't surprised a bit it blew up. What it is, they want to have production without any cost at all, and they cut 'em, they shaved 'em a little too thin."

"They're just like any hoggish corporation," said Clifford Walker. "They're fully covered on insurance and naturally they didn't pay too much attention on safety as far as I could see, because they don't stand to lose. People are to use, and they'll just use us, and that's the way it always has been and will be."

The Walkers were bitter, and understandably so. Days of pain and anguish had replaced the dream of a happy settled family life that had inspired Walker to bring his wife home with him to Georgia. Now totally disabled, Josephine Walker suffered from "terrific headaches" and sudden blackouts. One hand was crippled. Her weight had dropped considerably. She was receiving psychiatric treatment, including electroshock and medication.

"I've always been a person that can work out and keep the house—I can't do it now," she said. It upset her. "It's been difficult for me because there's a lot I can't do that I'd like to do." But it had been hard on her family too, and that was what worried her most. "I know, my husband being there, seeing me and carrying me to the hospital, he was in shock like I was, and my children were too." Her eight-year-old son visited her in the hospital. "I remember him sitting there and looking at me, tears coming out of his eyes, and he couldn't talk. It's—not been easy."

She was receiving thirty-seven dollars a week under Georgia workmen's compensation law. The payments would end after four hundred weeks. Mrs. Walker was one of at least fifteen disaster victims who intended to file suits against the federal government charging negligence in the disaster. Technically, as contractor for the trip flare, the government is a "third party" to the accident. Ideally lawyers would rather sue the company, but workmen's compensation laws, while insuring workers a sort of "no fault" if niggardly remedy, also deny the common law right to sue the employer. He is liable only for the amount of workmen's compensation—and usually, as is the case in Georgia, that amount is so small that it does more to protect the employer than the employee. "Third party" suits, then, are the only remedy available to workers.

Clifford Walker was an active member of the union, Local 832 of the International Chemical Workers. He was once vice president of the local. He said the union was weak because of racial divisions, inexperienced workers, and fast turnover. "People that works for $1.60 an hour, you're dealing with a bottom class of people." He believed building a strong union there would take several years and "a long training program." Because no one would expect a strong union organization in such a situation, and because the workers were part of a federal training program, he felt that the government should have been "more responsible."

The Georgia area office of the Manpower Administration, administered by the U. S. Department of Labor, was headquartered in Atlanta. Larry Giles, area supervisor in Atlanta, said he was "not really surprised" if, as workers claimed, Thiokol violated its Manpower contract by not providing adequate training or safety indoctrination. "It's been a problem with this particular program. There have been cases in which, although the company would sign the contract to perform these functions, they would in fact not perform them. Many of them just became cases where they put him on the production line and if he could perform, fine, if not, they would terminate him." And if safety should be left out? Well, "I guess safety is something we sort of take for granted in these things." So much for protection from the Manpower Administration.

And how about the Bureau of Labor Standards? In April 1971, just a few months after the disaster, the Occupational Safety and Health Act of 1970 went into effect, extending Labor Department health and safety jurisdiction to practically all places of work, with powers far exceeding those of the Walsh-Healey Act. Signing the bill on December 29, 1970, President Nixon praised it as legislation "that is going to do so much good for so many people across this land." The Bureau of Labor Standards became the Occupational Safety and Health Administration, and though the top officials remained substantially the same, labor leaders who had lobbied for the new act hoped the new agency would be more vigilant. Under the old Walsh-Healey Act the government could do little to penalize offending companies short of taking away their government contracts, but perhaps with the new, stiff penalties, and the power to shut down an "imminently dangerous" workplace, the government would not be so hesitant to enforce the law.

If OSHA is a more vigilant protection agency, no signs of it could be seen in the agency's handling of the Thiokol disaster. When I checked OSHA files in the Atlanta regional office a year and a half after the disaster, there was no record that Labor inspectors had been back to the Woodbine plant since the explosion. And OSHA officials seemed no more eager to help than they were as Bureau of Labor Standards officials. When I asked R. A. Wendell, assistant regional administrator, whether there had been more recent inspections, he said, "Well, you'll have to go to the Washington office for an answer to that. Washington is the office of record."

I sighed. "I thought Atlanta was the office of record."

"Well, actually, Savannah would be the office of record."

"Savannah?"

"Yes, Savannah is the area office. We have ten area offices."

And, what of OSHA's "plans for follow-up," the plans which were to "revolve about standards and regulations which may be developed from an analysis of the causes of the accident?" Well, they were working on that.

In Woodbine, a disaster victim wrote to her lawyer: "It may never happen again, but then again, it might. We pray that someone will be able to help us, and the people that will be working at Thiokol." But was anyone in Washington—or Atlanta—or Savannah listening?

THE NATION

The Crime of Greed

> Perfection of means and confusion of goals seem, in my opinion, to characterize our age. If we desire sincerely and passionately the safety, the welfare and the free development of the talents of men, we shall not be in want of the means to approach such a state.
>
> —*Albert Einstein*

"The truth," wrote Alice Hamilton in 1943, "is that the National Association of Manufacturers has fought the passage of occupational-disease compensation as it has fought laws against child labor, laws establishing a minimum wage for women and a maximum working day."

She continued: "Perhaps it is our instinctive American lawlessness that prompts us to oppose all legal control, even when we are willing to do of our own accord what the law requires. But I believe it is more the survival in American industry of the feudalistic spirit, for democracy has never yet really penetrated into many of our greatest industries."

In 1968, when an occupational health and safety bill was introduced in the House of Representatives, the National Association of Manufacturers did indeed oppose it. Said a spokesman: "The human factor is one of the most important causal elements involved in any accidental occurrence . . . Each employee must be motivated through training, education and supervision to understand and to want to perform work safely. This desire must come from within—it cannot be imposed through the threat of civil or criminal sanctions against the employer." And the NAM had been educating its workers, the spokesman pointed out, as far back as 1912, when it produced an educational film entitled *The Crime of Carelessness*. Instead of federal legislation, the NAM argued, it would be better to leave the law within the jurisdiction of the states—where, one might infer, there was virtually no law at all.

The Chamber of Commerce made the same argument, and pointed out: "The fact is, that the average American is safer at a workplace than he is at home, on the highway, or at play." And the Chamber cited the National Safety Council statistics so familiar and so often quoted because no others were available:

	DEATHS	DISABLING INJURIES
Motor vehicle	53,000	1,900,000
Home	28,500	4,300,000
Public	19,500	2,400,000
Work	14,200	2,200,000
(nonmotor vehicle)	11,200	2,100,000
(motor vehicle)	3,000	100,000

Statistics can lie, and the National Safety Council statistics distort conveniently for manufacturers. Supposing the statistics were accurate, which they are not, they still do not support the claim that the worker is safer at work than he is in his own home. An average 30,000 deaths at home must be considered in terms of 200,000,000 Americans, while the roughly 15,000 deaths in industry are for a much smaller population of 80 million workers. That is an incidence at home of approximately one death for every 7,000 people and an incidence in the factory of approximately one death for every 5,000 workers—in other words, the worker is, by those figures, considerably safer at home than he is at the plant.

The statistics lie in other ways. For example, they neglect to consider occupational diseases at all, and the best indications are that the figures for occupational disease far outweigh the accident statistics both in number and severity. Industry has defined the issue narrowly and comfortably to be one of industrial *accidents* and then shoved the burden of fault onto the worker. Once the issue is thus defined—not as a matter of deliberate skimping on safety for the sake of company profits but as a matter of careless workers—the problem becomes one of devising worker-incentive programs, safety awards, slogans, lights that flash when a "lost-time accident" occurs. It all can be done with a great display of righteousness. A casual visitor to a plant can't help but notice the lost-time scoreboard, the signs plastered everywhere, the safety trophies on the wall. And having devised the incentives—the awards, the newspaper announcements, the banquet presentations—for the best safety programs, corporations find ever more ingenious ways to protect their record, i.e. to suppress the true number of lost-time accidents. Who can be surprised if no one counted the injured worker crying in pain on a stretcher in a dark corner of the Mobil Oil refinery? After all, he remained on the premises. Or the woman folding towels from a wheelchair in the women's restroom at Chrysler, or the construction worker directing traffic with a bandage-swathed hand, or the many more like them? A worker is out eight weeks with a broken back? That will never do. Fire him for excessive absenteeism. Is the lost-time figure still too high? Well, fudge the figure, who will ever know?

With the Chamber of Commerce lauding the safety of the American working place and the NAM entrenched once again to resist, the 1968 occupational health and safety bill died in the House Rules Committee. Yet when it was revived in the next Congress, business groups, including the NAM and the Chamber of Commerce, testified in *support* of federal legislation. A sudden change of heart? Not exactly. Congressman William Steiger, who led the Republican forces on the Select Subcommittee on Labor, told me quite candidly: "I recognized that there was a problem—the amount of deaths—you couldn't deny that the problem existed. Once the need is demonstrated, the next question is what form should it take. The argument I made to the Chamber (of Commerce) was, 'Either you come out for a bill, or you're going to get the Daniels bill, and you don't want that—at least I think you don't want that.'"

The Chamber clearly did not want the Daniels bill—the labor-supported Democratic proposal.* The Daniels bill, for one thing, provided numerous "worker rights," a concept almost unheard of in the traditionally pro-management field of job safety. The most notorious worker right in the bill was the "strike with pay" provision which allowed workers to walk off the job under hazardous conditions. The provision, conceived by Gary Sellers, an aide of Ralph Nader and one of the key architects of the bill, was deleted in committee. Even labor opposed it because, as one labor lobbyist said, "It could create chaos." But there were many other provisions for workers' rights which labor strongly supported: the right for workers to know the results of tests of pollutants in the plant; the right to accompany a health and safety inspector during his rounds through the plant; the right to request a federal inspection; the right to medical records when exposed to toxic materials; and many others.

It was not at all evident that legislation must pass in 1970.

---

* Named for Dominick Daniels, chairman of the House subcommittee.

Most health and safety legislation in this country has been the result of major disaster and subsequent public outrage. The Coal Mine Health and Safety Act of 1969, for example, received impetus from an explosion and fire in a Consolidation Coal mine in Farmington, West Virginia, which killed seventy-eight men. Workmen's compensation laws were first introduced in most states between 1912 and 1920 after a series of well-publicized industrial disasters. Compensation acts were not extended to provide coverage of occupational diseases until the late 1930s, after exposure of the silicosis epidemic at Gauley Bridge. In 1970, there was no such single impetus for occupational safety and health legislation. One Congressman remarked that the political necessity for passing a bill was probably more apparent than real.

There were, however, significant reasons. Air and water pollution had been a major issue and resulted in legislation. The Coal Mine Health and Safety Act had been passed the year before. These and related issues and successes had helped to develop a political awareness in consumer groups, unions, and the Congress. And, as one Republican aide told me, "The administration wants a bill. That blue-collar strategy—they haven't given it up. The social issue—get the middle-class working man to vote Republican. The President wants that bill. He wanted it before the election to help Congressmen out. You can be damn sure he wants it now. If this thing is hanging in the fire in '72, he's in trouble. He wants the labor-management issue resolved."

The legislation that finally emerged was close to the Daniels bill. It is doubtful that the administration or business lobbies expected such strong legislation—they certainly did everything in their power to weaken key provisions—but labor committees in both the House and Senate were strongly pro-labor, and when a relatively weak House bill and a relatively stronger Senate bill reached conference, Democratic legislators yielded only on a few points. The key Republican Congressman, William Steiger, nevertheless seemed satisfied with those concessions and approved the conference report. It passed both houses and was signed into law by President Nixon on December 29, 1969.

The new law was named the Williams-Steiger Act after Senator Harrison Williams of New Jersey, chairman of the Senate Labor Committee, and Congressman Steiger. Workers had won unprecedented control over their environment in the form of worker rights. The act also provided a general duty clause—specific and tough language requiring that the Secretary of Labor shall set the standard for toxic substances "which most adequately assures, to the extent feasible, on the basis of the best available evidence, that no employee will suffer material impairment of health or functional capacity even if such employee has regular exposure to the hazard dealt with by such standard for the period of his working life." The act provided for the setting of emergency standards. There were provisions for citations and penalties of up to one thousand dollars for a serious violation of health or safety standards and up to ten thousand dollars for a willful violation. There was a crucial provision for establishing a National Institute for Occupational Safety and Health under the Department of Health, Education and Welfare. NIOSH would be responsible for development of the manpower—the training of researchers, scientists, and other professionals so crucial to carrying out the intent of the act. NIOSH also would develop the scientific criteria for health standards; it would make hazard evaluations of particular plants on requests; it would begin the long-neglected task of discovering health problems through industry-wide surveys; and it would conduct research on those problems.

President Nixon, in signing the bill, called it "perhaps one of the most important pieces of legislation to pass in this Congress . . . probably one of the most important pieces of legislation, from the standpoint of fifty-five million people who will be covered by it, ever passed by the Congress of the United States, because it involves their lives."

The few years that followed, however, proved that the Nixon

Administration's intent was, if anything, to delay progress rather than to advance it. It was as a Republican aide had predicted. "Even if the bill passes," he had said to me, "it will be tokenism. It won't be funded." As it turned out, the appropriations for job safety came to only one dollar per worker.

It was, in short, a good law rendered ineffective by lack of money, and in early 1974 it appeared that appropriations would remain low. Senators and Congressmen had been swamped with mail from owners of small businesses who had written to protest the Act as an unfair and sudden burden. John N. (Happy) Camp, a representative from Oklahoma, testified in House oversight hearings that businessmen in his district complained that "they have made alterations and spent a lot of money—but still have no way of knowing if they are in compliance . . . Small businesses simply do not have any way to go through the thousands upon thousands of words and regulations to try and find out how and why their businesses are not in compliance with the OSHA (Occupational Health and Safety Administration). Witness, for example, the seven pages of fine print published in the Federal Register as a result of the Act on stepladders alone!"

Legislation has twice passed both the House and the Senate which would exempt businesses employing fifteen workers or fewer. The legislation was attached to appropriations bills subsequently vetoed by President Nixon for other reasons. If signed into law, the bills would have excluded from coverage an estimated eighty-six percent of all employers and approximately thirty percent of all workers.

In January 1974 one spokeswoman on the House Labor Committee staff reported: "There are seventy-six bills before us now to amend it and three to repeal it." The House Appropriations Committee has felt the pressure too. The committee cut funding for additional Labor Department inspectors from 723 requested by the Nixon Administration to 691. (After conference with the Senate, which had authorized 1,100, the figure was compromised at 800 inspectors). "They were lucky to get away with that," said a committee spokesman. "It's a pretty hot potato. Many members of this subcommittee would like to gut it."

The "knot of the problem," George Taylor of the AFL-CIO told the House Select Labor Subcommittee, is not the law, but the Department of Labor. "From the outset, OSHA has failed to provide and distribute to all covered employments clear, understandable guidelines to the standards . . . It is yet to accomplish this by means of a massive informational program directed to every covered employer. As a result, thousands of small employers are either ignorant or frightened of OSHA."

George Guenther, the Undersecretary of Labor for Occupational Safety and Health, told the same committee that such a mailing would be too expensive, and he made it clear that he wasn't interested in informing employers more adequately anyway. "At the present time," he told the Congressmen, "we believe we have adequate personnel and financial resources to carry on the program."

What OSHA had done to inform employers and employees, besides answer inquiries, make speeches, conduct seminars, and distribute press releases and various publications, was to release six thousand radio and eight hundred TV commercials for broadcast as public service announcements. A worker hearing one of these spots might not be aware that it had anything to do with a new law let alone worker rights. There just wasn't time in a thirty-second announcement, Labor Department spokesmen claimed, to mention the act. There was time, however, to pound home to workers the message that industry has been preaching for years—that they'd live a lot longer if they'd only watch their step:

DEPARTMENT OF LABOR
RADIO PSA   :10 SECONDS
Celeb: (With music background) This is Kenny Price. The
            U. S. Labor Department says that slipping and falling

accidents on the job can be avoided. Pay attention. Think safety while you work.

DEPARTMENT OF LABOR
RADIO PSA   :30 SECONDS

SOUND: APPLAUSE & SFX OF "DINNER"

Anncr: That's old Andy—finally getting his gold watch after 31 years on the line. Good worker—safe, too. Not the kind of guy who'd put safety last. Maybe that's why he made it—accident free for 31 years. Andy knows Job Safety really pays.

SOUND: "CONVERSATION" BECOMES LOUDER . . . THEN FADES

Anncr: Practice job safety all the time. For more information, contact the nearest office of the U. S. Department of Labor.

DEPARTMENT OF LABOR
RADIO PSA   :20 SECONDS

SOUND: FACTORY "QUITTING" WHISTLE

Barney: That's the quittin' whistle, Fred. Let's go home.

Fred: Better wipe off that grinding wheel, Barney boy. Remember—safety first!

Barney: That's right, Fred. When we follow the job safety rules, the next shift will be in good shape too. Job safety is everybody's job. For more information, contact the nearest office of the U. S. Department of Labor.

Perhaps the most damaging blow dealt the legislation by the Nixon Administration was the immediate decision to hand enforcement back over to the states. The law provides for such action, but as Jack Sheehan, a Steelworker union lobbyist pointed out, the federal law was enacted "in response to the historic *failure* of the states to provide adequate protection to the working man." Under the federal law, states could submit plans for their own legislation, which must be "as effective as" federal law. States which had not had plans approved by December 1972 would be preempted by federal jurisdiction. At that date, however, only thirteen states had submitted plans. Guenther then proposed a six-month extension of the deadline. The AFL-CIO obtained a court order preventing the extension.

Organized labor has since opposed all state plans. Sheldon Samuels of the AFL-CIO, a career environmentalist who worked in state and federal government during the Johnson Administration, said that one of the things labor had learned, during that period, is that "some things are best done by the federal government, some things are best done by the states." He said states are best equipped to provide technical services and educational programs and so on. But the states should not be entrusted with enforcement, he said, or the setting of standards. "You can't have fifty sets of permit systems to control carcinogens, for example, and have it administratively viable. That's best administered by the federal government."

The administration has proceeded reluctantly in the establishment of standards. By January 1974 it had promulgated only one new standard and that one, for asbestos, was the result of considerable prodding by labor and public interest groups. OSHA's laggardliness is deliberate. An official of a private standards-setting body summarized an OSHA memo he had seen: "The OSHA standards package will not be augmented to any extensive degree in the near future . . . OSHA will move with great deliberation in the standards area, and would prefer to see industry adopt and follow its own voluntary standards. This practice may lead to governmental control only in principal problems areas."

The policies of the administration were summed up accurately by Bob Kasen, editor of the Teamsters publication, *Focus*: "Within OSHA, itself, there has been no sense of mission or dedication to the implementation of the Act on behalf of the workers it was intended to protect. Rather, there is a steady and prevailing attitude in the top echelons of OSHA that shows priority compassion for problems and inconveniences of management."

Such compassion has been most apparent in OSHA enforcement policies. The department has estimated that five million business establishments are covered by the act, yet in January 1974, OSHA employed only 579 compliance officers. In 1972 the Department of Labor made a total of 36,100 inspections, found 125,400 "alleged violations," and issued 23,900 citations with penalties totaling $3,121,000. That averages out to a trifling twenty-five dollars per violation. Unions have complained that requests for inspections have resulted in long delays. A Steelworkers survey revealed delays in 1972 of from two weeks to three months before OSHA responded to safety complaints and as much as one year for health complaints which would require an inspection by an industrial hygienist.

The National Institute for Occupational Safety and Health has similarly suffered under the Nixon Administration. The federal occupational health effort, traditionally underfunded, understaffed, and politically impotent, has taken just a tiny step forward, with a 1974 appropriation of only twenty-eight million dollars, ten million dollars of which is earmarked for health activities under the Coal Mine Safety Act.

There have been periods— the early 1900s and again in the '30s and '40s—when public health officials, in Dr. Harriet Hardy's words, "did a bang-up job." Edward Baier, director of the Pennsylvania occupational health program, labeled the period just before, during, and after World War II as the "golden age of occupational health." Public attention shifted to radiological health and air pollution in the late '40s and '50s, he said, and funds were also shifted. "Attrition of personnel from the ranks of occupational health to those related fields followed and unfortunately spelled the demise of many state programs."

Dr. Marcus M. Key, the mild-mannered director of the new NIOSH and former administrator of the old Bureau of Occupational Safety and Health has rarely shown the political toughness necessary to run a strong program. Perhaps he believed it necessary to bend with the wind in order to survive at all—but survive for what? Key's public statements have been on the whole more aggressive since the establishment of the Institute, although in reply to a query at the 1972 House oversight hearings regarding the Institute's budget, Key said that it was adequate. "Considering the budget restraints and demands of other programs," he said, "I think we are getting a fair share."

At an April 1973 meeting of the National Advisory Committee on Occupational Safety and Health, a member of the committee suggested to Dr. Key that NIOSH do more research in the field of safety. To my surprise, Key finally got mad. "Nothing irritates me more than being asked to do more for less," he snapped. Although the budget had been increased slightly, he explained the number of positions would go down. "NIOSH, I don't think, is a viable organization at this time," he said.

Dr. Joseph Wagoner and Dr. William Johnson, NIOSH officials, gave me a good description of the Institute's problems when I visited the NIOSH offices on Broadway in Cincinnati in March 1973. "When I was coming to Cincinnati," said Wagoner, "I was told if 1014 Broadway wasn't there nobody would know the difference. I was told it was a good setup, you could play pool, or that you could pick up forty thousand dollars a year hitting the emergency room at night, sleep on a cot during the day. Some of this is still going on."

"Professional standards within this agency are something else," added Johnson. "Talk about what we inherited." He listed the chief of the field studies branch, who, he said, spent four days a week attending medical school; the assistant chief, who was attending law school three nights a week; and the acting chief of biometry, who was studying for his master's degree.

While the old problems of incompetence and conflicts of interest seem to be disappearing, plenty of others have developed to take their place. Public health services have in the past been staffed largely by physicians and health professionals fulfilling their military duty, but with the end of the draft, said Johnson, "the physician pool for health services is drying up completely." Attracting other qualified people is a major problem under HEW budget restrictions.

Under the leadership of competent and dedicated young men like Wagoner and Johnson, the Institute is finally beginning once again to do a "bang-up job." A series of industry studies have begun and data is coming out at an impressive rate, but there is far more to do—and far too little with which to do it.

In a 1972 HEW publication Dr. Marcus Key was quoted as saying: "The effectiveness of the Occupational Safety and Health Act of 1970 is directly dependent upon NIOSH's ability to produce sufficient manpower to carry it out. That represents our greatest challenge."

The act authorizes HEW to carry out programs "directly, or by grants and contracts," to produce the necessary personnel "to carry out the purposes of the act." HEW has estimated that to provide the necessary manpower, thirty-five to forty-five thousand additional personnel would need to be trained, including ten thousand industrial hygienists, ten thousand occupational health nurses, three thousand occupational health physicians, five to ten thousand safety professionals, and eight thousand occupational health scientists "of various types."

One of the most effective measures of the government's commitment to a program is the program's provision for manpower training. There will never be an adequate occupational health program in this country until there are enough health professionals to run the programs. In the 1950s, when the government began to develop atomic energy and saw a future for peaceful use of this power source, enormous amounts of federal money were made available to medical and public health schools to train personnel. Whole departments of radiation health were established in public health schools and still exist. Yet in occupational health, the 1974 fiscal budget—for the first year since 1937—provided no funds whatsoever for manpower training. The only other federal source for such funds would be the Comprehensive Health Manpower Training Act of 1971, which provided funds for physician training and public health training. A health crisis developed in 1973 when President Nixon impounded those funds, which medical and public health schools depended on for many of their graduate programs. The funds were eventually released and the 1974 budget was an increase over the former year according to an HEW spokesman, but he said no specific money was available for occupational health. "It would be very low on their list of priorities."

This dismantling of the national health manpower effort, which effort the Act so clearly sought to encourage, has been the most insidious decision of the administration to date. By providing no trained people for the future, it would make it impossible for years to come to put the Act to full use and do a good job of protecting workers.

One might well ask where the unions have been all these years, as conditions in the plants have been declining. The answer is that until recently they have not been at the forefront of any battle for better working conditions. The reasons are many and, I believe, strongly related to more general problems of unions. Most have become bureaucratized until the interests of workers are not the interests of union leaders. Leaders have become so remote that they are not even aware of the problems of their constituents.

In the last few years a few unions have been fighting for better plant conditions. The outstanding example is the Oil, Chemical and Atomic Workers Union (OCAW), a small union which I relied on heavily for information about plants because no other union, with the exception of the Textile Workers Union of America, was able to give me information about conditions at specific plants. Tony Mazzocchi of the OCAW has said he views the health and safety law as merely a tool. The real battle must be waged, he said "by the workers themselves, at the point of production." The OCAW has been successful in negotiating for sophisticated health and safety provisions in their contracts with oil companies. In 1973 the UAW, under pressure from dissident groups within the union for several years, finally followed suit in agreements with the auto industry.

Until such a time as an administration interested in vigorous enforcement of the health and safety legislation emerges, such efforts on the part of labor are the best hope for change. As Tony Mazzocchi said in Congressional testimony: "We have lived a long time under many administrations. We know that in the final analysis the government will never do the type of job that has to be done because of all the facts and forces that exist, as well as because of the society in which we live . . . We are going to have to depend on what we do ourselves through collective bargaining.

"Of course," he added, "this excludes seventy-five percent of the people who work for wages in this country (who are unorganized). We recognize that. We do hold stewardship for those people by virtue of saying this is what is wrong, fellows and ladies, and the only way you are going to correct the situation is to organize. The government is not going to do it for you. The interpretation of the act mitigates against the government helping the unorganized. He is the forgotten man."

CONCLUSION

The Savage Cycle

There are some things for which there can be no justification, and torture is one. Its evil is infectious, corrupting not only the torturer but those who try to pass by on the other side . . . The Talmud explained why 1,500 years ago: "Therefore was a single man only first created, to teach thee that whosoever destroys a single soul from the children of man, Scripture charges him as though he had destroyed the whole world."

—Anthony Lewis, *The New York Times*

The plants I visited during the three and a half years I worked on this book were owned and operated for the most part by huge corporations: Mobil Oil, Union Carbide, Chrysler, Ford Motor, Thiokol, Anaconda, Bethlehem Steel, Minnesota Mining—cor-

292

porations with the resources to measure and control industrial hazards. Yet now as throughout American history, companies such as these shrug at the pleas of workers whose health they destroy in order to save money. They hire experts—physicians and researchers—who purposely misdiagnose industrial diseases as the ordinary diseases of life, write biased reports, and divert research from vital questions. They fight against regulation as unnecessary and cry that it will bring ruination. They ravage the people as they have the land, causing millions to suffer needlessly and hundreds of thousands to die.

This slaughter is industry's cardinal secret, hidden from public scrutiny beyond plant gates, guarded jealously, supposedly to protect "trade secrets" and for "security reasons." (Is it less than coincidence that the plant safety director is often in charge of security as well?) It is hidden too beneath the lies of the National Safety Council, which tells us that job safety is the worker's responsibility, just as it has told us that the key to auto safety is fastening our seat belts. The truth has been hidden so successfully for so long perhaps because if the truth were fully known, the American profit system would be shaken to the roots.

Men at the top of corporations have repressed the horror of what they have wrought and concerned themselves more intimately with the profit and loss statements, mergers, acquisitions and politics of corporate power. Plant managers seem likewise preoccupied. One industrial hygienist I know was surveying a plant when he noticed the plant manager running through the plant, from one checkpoint to another. "He knew what production had to be every hour in order to make production that day," said the engineer. "If he didn't make production that day, he wouldn't make it that week, and if he didn't make it that week, he might not make it for the month, and he might not get his bonus at the end of the year."

Corporations devise the incentives that make the system work the way it does. What goes on in the minds of the men who conceive and carry out such policies? What rationalizations must they make? How can they believe the lies they must tell? It is likely that corporate management has relied on the optimistic reports of subordinates and professionals who were willing to say what they knew management wanted to hear. Once started down the trail of half-truths, it is hard to turn back, repudiating one's past decisions and beliefs—and few have turned back. I have heard of only one who did. The same industrial hygienist who told me the story above was once asked to test for mercury exposure in a Puerto Rican plant employing primarily young Puerto Rican women. When he presented the results to the plant's manager, he said, and explained that the mercury levels were so high that many of the women must certainly have been poisoned, the manager wept and shortly afterward resigned.

Frequently no one higher than the plant safety director is concerned with the problems, and he seldom has the knowledge, let alone the power, to protect the workers. Michael Krikorian, president of the American Society of Safety Engineers, interviewed in the December 1973 issue of *Occupational Hazards,* a journal for safety professionals, was quoted as saying that safety directors often don't have the first idea what health hazards exist in the plant. "They go around," said Krikorian, "with a little fear gnawing away at the backs of their minds that one day they'll find themselves faced with a full-blown crisis, all the while telling themselves it can't happen here."

"He reminds safety directors," said the magazine: " 'It has "happened here" to companies dealing with asbestos, carcinogens, spray adhesives, and most recently at the Columbus Coated Fabrics plant of Borden Co.' "

In February 1974 B. F. Goodrich announced it suspected that the deaths of four rubber workers, killed by a rare type of cancer of the liver, were connected to exposure to the chemical vinyl chloride, used in the production of polyvinyl chloride. Nineteen companies make the chemical—some five billion pounds of it were produced in 1973, according to a *Wall Street Journal* article—and its use is widespread throughout the plastics industry. Such reports are not surprising when one considers that several hundred thousand chemicals have been dumped virtually unknown and uncontrolled into the industrial environment. Cancers from industrial exposures usually take twenty to thirty years to develop. And since the majority of chemicals in use today have been created only in the last twenty to thirty years, these exposures can be predicted to result in many such tragedies.

Until research is done to determine whether these cancers are occupationally induced, workers exposed to petrochemicals can only wonder at the number of strange diseases that afflict them. Lou Mariani, who worked in the lab at Texaco, said that technicians there were exposed to cumene, benzene, and toluene and that ventilation in the lab was poor. "We started our campaign to get better ventilation in that lab fifteen years ago," he said. "We've had no success. In the meantime, we've had in the lab two cases of men with blood disorders—blood vessels eaten away and plastic had to be put in. Two more men, a technician and a tester, had blood disorders. In 1963 we started hollering for blood tests and rotation of men. In my case, they notified me of an irregularity two years ago. I went to my own doctor and he found me to be okay. We did have a guy, the head of the lab, who died of leukemia. That was in 1966 or 1967. One man died from spinal cord degeneration. They say another man has the same thing. There's another guy with a circulatory problem. He's been out seven months. They don't know—none of these cases have been considered work-related and there's no way of proving it."

It is frightening to hear stories like this and to suspect that they are being repeated in thousands of factories across the country. Why must millions of Americans work at the mercy of manufacturers who refuse to take into account the toxic effect the chemicals and minerals may have on the workers who must use and produce them?

Mass production has brought us goods at cheaper prices only if we do not count the price paid by the men and women who have suffered in the process. In 1906 Edwin Markham wrote a moving description of child labor in American cotton mills. "To what purpose," he concluded, "is our 'age of enlightenment,' if, just to cover our nakedness, we establish among us a barbarism that overshadows the barbarism of the savage cycle? . . . Is this what our orators mean when they jubilate over 'civilization' and 'the progress of the species'?"

Only the American people themselves have the power to bring about the changes that can stop this industrial massacre. When will they demand the rights that industry has done so much to subvert, including the most basic right of all—the right to life? Workers cannot do it alone. They need the help and support of all of us. I hope it is not long before a majority of American people recognize the problems that today are understood well by relatively few.

One who did understand, all too well, was John Niewierowski, a worker at FMC's American Viscose rayon mill in Nitro, West Virginia. Niewierowski, described earlier in this book as having been poisoned by carbon disulfide while working as a cutterman, said he was concerned about the conditions at the plant but felt powerless to change them. "We have trouble," he told me. "We're kind of like a lost boy in the high weeds. We know we want out, but we just can't get any help. The American worker everywhere is supporting society," he said. "The American society, they owe the worker something."

INDEX